VON RICHTHOFEN

THE LEGEND EVALUATED

RICHARD TOWNSHEND BICKERS

Naval Institute Press
Annapolis, Maryland

First published in the United Kingdom in 1996 by Airlife Publishing Ltd.

Published and distributed in the United States of America and Canada by the Naval Institute Press, 118 Maryland Avenue, Annapolis, MD 21402-5035.

Library of Congress Catalog Card Number:
96-70935

ISBN 1-55750-571-3

This edition is authorized for sale only in the United States of America, its territories and possessions, and Canada.

Printed in Great Britain on acid-free paper

CONTENTS

FOREWORD

Rittmeister Manfred *Freiherr* von Richthofen is a dashing and aristocratic designation that loses nothing in translation: Cavalry Captain Baron Manfred von Richthofen. It well befits the character and short but brilliant life of the man who bore it.

His autobiography, *Der Rote Kampfflieger*, was published in Germany in 1917, many months before his death in action the following year; in America as *Red Battle Flyer* in 1928 and in France as *Le Corsaire Rouge* in 1931. He has had German, British and American biographers. He also figures in numerous books on general air operations in the First World War or specifically about fighter aces. Several have been translated from the original languages for foreign publication.

Despite so much attention, he remains not only a complex but also an enigmatic and controversial personality and there is still something fresh to say about him.

He is generally viewed with respect and admiration. The most jaundiced exception is Arch Whitehouse, according to whom,

> 'Aggressive though he was, von Richthofen made more than exaggerated claims and never, if he could avoid it, gave credit to a comrade for victory. He was notably jealous of his brother Lothar and disgustingly arrogant before young pilots of his *Staffel*.'

But Whitehouse's sour account of his own service on the Western Front as a cavalry trooper and Royal Flying Corps air gunner makes a heavy demand on one's credulity and leaves an unattractive impression of the writer rather than his subject.

1

This present work is an evaluation of Richthofen as a fighter pilot and leader in comparison with illustrious predecessors, contemporaries and successors of his own and other nationalities; and his influence on air fighting from his own time to the Korean War, after which homing missiles rather than guns have predominated.

The basic principles for a fighter pilot emanate from the brief dictum of *Hauptmann* Oswald Boelcke, the first great fighter innovator and leader, in 1915. In answer to Richthofen's question about what accounted for his outstanding success, he replied, 'Well, it's quite simple. I fly close to my man and aim well, and then of course he falls down.' This simplistic statement might have been merely a dismissive or jocular retort, not a concise definition, but it nevertheless expresses the essence of the matter.

In 1917, *Leutnant* Carl Degelow, who was to be the last recipient of Germany's highest decoration, the *Pour le Mérite*, confirmed that the essence of victory was 'To get ever closer, metre by metre – even three or four metres.' Such proximity suggests a certain rashness: if the enemy aircraft exploded, the attacker could also be blown to bits or his own machine hurled out of control to an almost certain crash. However, caution has never been conspicuous in a fighter pilot's nature.

In 1939, Group Captain Sir Douglas Bader KBE, DSO, DFC, one of the greatest fighter pilots and leaders of the Second World War, based his tactics on the methods of the 1914–1918 British pilots of whom he had read as a schoolboy, and cadet at the RAF College, Cranwell. 'Get as high above your enemy as you can, with the sun behind you, and you'll have the advantage. Shoot from as close as possible.'

Group Captain A.G. 'Sailor' Malan DSO, DFC, whom many of his most distinguished comrades have rated highest among World War Two fighter leaders, set out the following 'Ten Commandments' in 1940.

1. Wait until you see the white of his eyes. Fire short bursts of one to two seconds, and only when your sights are definitely 'on'.
2. Whilst shooting think of nothing else. Brace the whole of

2

the body, have both hands on the stick, concentrate on your ring sight.

3. Always keep a sharp look-out, 'keep your finger out'.
4. Height gives you the initiative.
5. Always turn and face the attack.
6. Make your decision promptly. It is better to act quickly, even though your tactics are not of the best.
7. Never fly straight and level for more than thirty seconds in the combat area.
8. When diving to attack, always leave a proportion of your formation above to act as top guard.
9. INITIATIVE, AGGRESSION, AIR DISCIPLINE and TEAM WORK are words that MEAN something in air fighting.
10. Go in quickly – Punch hard – Get out!

Had Richthofen been alive then, by which time weaponry had been much improved, and aircraft performance and the size of fighter and bomber formations had greatly increased, we can be sure that he would have given the same advice as Bader and Malan. He was already practising and teaching some of these principles.

To assess his skill as a fighter pilot in comparison with his contemporaries and those of later generations, there is the evidence of his record and the types of enemy aircraft he destroyed. To gauge his influence over his comrades, the verbal comments, books, articles and letters by some of them are available. To judge his ability as a squadron or Wing leader and his influence on air fighting and leadership in later wars, he must also be examined for a more profound and rare gift. In his brilliant book, *The Ace Factor*, Mike Spick identifies this as 'The ability of the pilot to keep track of events and foresee occurrences in the fast-moving, dynamic scenario of air warfare', known now as Situational Awareness.

This is sometimes a natural talent and sometimes the fruit of experience. Whichever way it is acquired, those who possess it have been the most effective fighter leaders: Manfred von Richthofen was one of them.

CHAPTER 1

TWO RECORDS IN THE RICHTHOFEN TRADITION

On 1 September 1942, *Oberleutnant* Hans-Joachim Marseille, commanding a Messerschmitt 109 squadron in the North African desert, gave a fine display of the example set by Manfred von Richthofen a quarter-century earlier.

On the night of 30 August, *General* Rommel, commanding the *Afrika Korps*, had launched the attack that he believed would finally defeat General Montgomery's Eighth Army. The land battle was supported on both sides by bombing. At 7.30 two mornings later, Marseille, leading the fighter escort, joined a formation of Ju87 dive bombers. There were no other aircraft in sight until, nearly an hour later, he warned his pilots that he had spotted the enemy approaching and above. 'Ten minute dots that swiftly grew bigger', the official report says.

The Messerschmitts were at 3,500 metres. As soon as the *Stukas* began their dive, Marseille put his machine into a climbing turn to the right and called, 'I'm attacking.'

The account continues,

'In a twinkling, his No. 2 saw him swing to the left, position himself behind the last of the suddenly scattering Curtiss P40s [Kittyhawks] and shoot at a range of 100 metres. The enemy aeroplane lurched abruptly on to its left wingtip and, burning, fell like a stone vertically to the ground. The time was 8.20.'

Like their predecessors in the First World War, German pilots' aircraft recognition was poor. So, apparently, was this lot's arithmetic. The aircraft were Hurricanes of No. 1 Squadron South African Air Force; and there were twelve of them.

The Ju87s had bombed, so the escort went down to stay with them. At 8.30, according to his No. 2, Marseille shot down another alleged Kittyhawk in flames. Three minutes later, by the No. 2's timekeeping, his leader despatched a third and 'the flames from burning wreckage lit the desert'.

While Marseille and his wing man had broken off to deal with the intruders, the rest of the escort had stayed with the bombers. One of them now gave another alarm: 'Spitfires!' This time the identification was correct. Six, belonging to 92 Squadron, swept down on the two 109s from above and astern. One Spitfire opened fire with cannon and machine-guns and tracer glittered past Marseille. He turned tightly, climbing; his opponent overshot and he was now on its tail. At 80 metres' range he shot it down. The time, said his winger, was 8.39.

What the German report omitted was that six Messerschmitts were damaged in this fight.

Allied documents show that twelve Hurricanes of 238 Squadron also took part in this action. Major Metelerkamp and Lieutenant Bailey of 1 Squadron SAAF were wounded and crash-landed their severely damaged aircraft. Flying Officer Matthews of 238 Squadron also crash-landed. Pilot Officer Bradley-Smith of 92 Squadron baled out. So five victories was the correct score.

Marseille's marksmanship was such that when he landed at 9.14 his armourer found that he had used only 80 rounds of cannon and 240 of machine-gun ammunition.

At 10.24 am he took off with two others to escort another *Stuka* raid. (The German records emphasise that there were only three and not the twelve attributed in some accounts.) When nearing the target, Marseille saw two Allied formations, each of fifteen to eighteen bombers and twenty-five to thirty Kittyhawks. Twelve of the latter belonged to No. 3 Squadron, Royal Australian Air Force.

The account of this operation superfluously informs us that 'He had never been a man to be impressed by superior numbers of the enemy.' There was not much point in being a fighter pilot if one were, as even the least experienced in RAF squadrons had demonstrated over southern England in the summer of 1940.

When eight Kittyhawks peeled off to attack the *Stukas*, Marseille turned his section of three towards them. The Kittyhawks formed a defensive circle. The German chronicler wrote, 'This tactic was effective in normal circumstances, but not against Marseille.' He positioned himself in the middle of the circle and picked off a Kittyhawk at 50 yards' range. His wing man noted that the time was 10.55. Half a minute later, he sent another down.

'Suddenly the circle fell apart. The leader lost his nerve and they all broke away to the north-west.' It should not have been taken for granted that the Kittyhawks' leader had been frightened off; it was part of his duty to protect his pilots' lives so that they could fight another day.

This was a futile move, however, for in two minutes Marseille overtook them and shot another down. The five survivors flew eastward with Marseille in pursuit. He soon scored again. The remaining four now headed north-west, found themselves over the sea, so returned landward. Two minutes elapsed before one of these was despatched. One minute later the sixth, and within three more minutes the last two, were shot down. His eight victories had taken ten minutes.

Generalfeldmarschall Kesselring, commanding *Luftflotte* (Air Fleet) 2, had come to visit the *Geschwader* (Wing) which comprised three *Staffeln*. When Marseille landed at 11.35 he informed Kesselring that his *Staffel* had shot down twelve aircraft that morning.

'How many did *you* get?'

'Twelve, *Herr Feldmarschall*'

Eyewitnesses relate that 'The *Generalfeldmarschall* shook the officer by the hand, then sat down lost for words'.

At 5 pm Marseille took off with his *Staffel* to escort Ju88 bombers. The *Luftwaffe* account of what happened on this operation alleges that fifteen fighters, which Marseille again misidentified as Kittyhawks, attacked. They were actually twelve Hurricanes of 213 Squadron.

The narrative goes on to state:

'. . . at heights between 1,500 and 100 metres, Marseille's first four victims fell at roughly one-minute intervals between 5.45 and 5.30, the fifth at 5.53. The places where they went down were 7km south, 8km south-east, 6km south-east, 9km south-south-east and 7km south-south-west of Imayid.'

Marseille's claims for the day therefore totalled seventeen.

Apparently the other pilots took little or no part in the fight. Whether Marseille ordered them to leave all the destruction to him while they guarded him, as he usually did, or whether he was so brilliant a shot and so fast in manoeuvre that he beat his comrades to the kill every time, is not revealed. A wing man's job was to protect his leader's tail and attack only when necessary, but with so many adversaries in so small a compass one would have expected him to do some shooting.

Even if none of them engaged the RAF fighters, they must all have been taking evasive action while the dogfight was going on. It therefore seems incredible that in the turmoil such accurate observation and timing could have been made of where and when victims fell, at five widely scattered pinpoints. If Marseille's wing man memorised them, he must have been unusually gifted; and, with one hand on the throttle and the other on the stick, busy keeping station on his leader, how could he have made written notes?

In the apposite words of Sir Boyle Roche in the House of Commons in the late Eighteenth Century: 'Mr Speaker, I smell a rat; I see him forming in the air and darkening the sky; but I'll nip him in the bud.'

The RAF archives confirm that 213 Squadron did lose five aeroplanes in this encounter, Flying Officer Wollaston was killed and Sergeant Potter was missing, presumed dead, Sergeant Garrod baled out, Flying Officer Avise and Flight Sergeant Ross crash-landed; all three unhurt.

Leutnant Stahlschmidt, flying in the *Staffel*, claimed two of them; so perhaps Marseille's tally was really 15.

The total Allied fighter losses over the desert that day were twenty, but the Germans claimed twenty-six. There is no implication of deliberate falsification. It is well known that accurate reckoning of aircraft shot down in an air battle between large numbers has always been difficult.

Asked, forty-six years later, for his evaluation of Marseille, *Generalleutnant* Adolf Galland, who commanded the *Luftwaffe* fighter force from November 1941 until two months before the war ended, replied without hesitation, 'He was the best.'

Nonetheless, although he was an inspiring *Staffel* commander and a miraculous shot, he was hardly the quintessential great fighter leader. Like many fighter pilots in all the air forces that fought in both world wars, he was more interested in adding to his score than in the worth of the aircraft he destroyed. In that characteristic he resembled Richthofen. Marseille tried to shoot down escorting fighters rather than cut through to the bombers, which should have been the priority targets for every conscientious fighter pilot: because in both wars the bomber was the prime instrument of aerial warfare and did the greatest damage.

There was a difference, however, in their attitude to achieving a high score. Fighter pilots of every generation have, according to their nature, fallen into two categories: hunters and shooters. The definition of this distinction is commonly attributed to Manfred, but it was actually his father who originated the terms in this context. By nature, Manfred was a hunter. Totally committed to the stalking and shooting down of an enemy aircraft when he saw one, until half-way through his career in fighters he was usually satisfied when he had done so. The shooters, such as his younger brother Lothar, whom he chided for this characteristic, rushed into an attack and were not content with the result of a sortie if they gained one victory and had enough fuel and ammunition left to do some more execution. But such is the insidious, heady effect of accumulating success on a grand scale, that by the time Manfred had destroyed forty British aircraft he had, he admitted, become as eager a shooter as Lothar, who started later than he and whose score was mounting rapidly; and sibling rivalry no doubt added to the elder's lust to kill.

Marseille had many opportunities to attack bomber formations escorted by fighters. Richthofen had as many to attack escorted reconnaissance machines; but from what can

9

be deduced about his character it seems probable that he would also have gone for quantity rather than quality in the circumstances common in the Second World War.

There were two basic differences between outstandingly successful fighter pilots, both hunters and destroyers; naturally gifted and therefore exceptional pilots who learned to shoot accurately, and born marksmen who learned to fly adequately. This distinction, however, has to be qualified. Throughout his career a military pilot is assessed from time to time in one of four categories, starting at elementary flying training school; below average, average, above average and exceptional. However, as his flying hours and operational experience increase, an average pilot might well earn higher assessments. Group Captain Sir Hugh 'Cocky' Dundas KBE, DSO and bar, DFC, described all the pilots on No. 56 Squadron when he commanded it in 1942, as above average.

On 11 December 1940, a pilot of No. 33 Squadron, RAF, flying a Hurricane, performed a feat that neither Marseille nor anyone else equalled in either world war. He shot down seven enemy aeroplanes in under fifteen seconds.

Pilot Officer Charles Harold Dyson DFC, known as 'Deadstick' because he had made more than one forced landing in the desert, owing to engine failure, was returning alone from a patrol over the Libyan Desert that morning. Emerging from a small cloud, he was astonished to see seven Italian aircraft close in front of him. One was a Savoia-Marchetti S59 bomber. Escorting it, in two Vs of three, were six Fiat CR42 fighters. The Italian fighter leader had not taken the precaution of detailing one of his pilots to weave astern of the formation and give warning of any approaching enemy aircraft. Nor evidently was the bomber's dorsal gunner, who was supposed to keep a sharp lookout astern, alert.

Dyson opened fire at once. His first burst set the three rearward fighters alight and his second took out the leading three. For the eight machine-guns of a Hurricane (or Spitfire) there was only 14.8 seconds' worth of ammunition. None was left with which to attack the bomber. Dyson watched the six fighters going down in flames, then nipped back into cloud in case there were any more hostile fighters around.

When he emerged, another six CR42s jumped him and he spent the next few minutes dodging their bullets until he could safely dive away at full throttle. His engine had been hit and failed, so he had to make another deadstick landing – 120 miles from base. He did not arrive back there until six days later, after much walking and lifts from Army vehicles. Only then did he learn that ground troops who had watched him destroy six Italian fighters had also seen one of these collide with the bomber, which caught alight and crashed. So he had scored seven victories in one action that lasted less than a quarter of a minute. He was awarded a bar to his DFC.

Richthofen, in Valhalla, must have envied him and admired the aggressive spirit and determination to press home an attack from close range and down to the last round, which matched his own. This, moreover, was a hunter's, not a shooter's, achievement; Dyson's multiple success was owed to the fact that he took all seven victims by surprise – he himself being considerably surprised, too, on seeing them – and not by attacking each one in turn with forethought.

Dyson scored two more victories in the desert before his squadron was posted to Greece. There, he was shot down twice, by a Fiat G50 fighter and by flak. He is estimated to have destroyed another sixteen enemy aeroplanes in Greece and Crete, but nearly all RAF records of that short campaign were lost. He ended it as a prisoner of war.

Chapter 2

Know Your Enemy

Manfred was officially credited with his first victory on 17 September 1916. He was flying an Albatros D2, armed with two forward-firing Spandau belt-fed machine-guns, each loaded with 300 rounds and synchronised to fire between the propeller blades. It was Germany's latest fighter and the most handsome military aeroplane in the world, sleek and torpedo-shaped. There is an old maxim in the aviation world, 'If it looks right, it will fly right'; the new fighter personified this. Its 160 hp Mercedes engine gave it a maximum speed of 109 mph; in six minutes it climbed to 3,280 ft, to 6,500 ft in 9 minutes and reached its ceiling of 16,400 ft in 37 minutes.

His victim was a two-seater fighter, the de Havilland FE2b, capable of 80 mph and taking 52 minutes to reach 10,000 ft. This was a pusher type, the engine and propeller directly behind the pilot, who had no gun. In a crash the engine was often jolted forward and crushed him. In front of him sat the observer, with a Lewis gun mounted on a post so that it could be swung through 180 degrees ahead as well as raised and lowered. It could also be aimed astern, if he kept it tilted high enough not to hit the upper wings or the propeller. Only bursts of not more than ten rounds could be fired, or the barrel would be ruined. Ammunition was contained in a drum of 47 rounds. Removing an empty drum and fitting a heavy full one was difficult: the observer had to stand, buffeted by the rush of air, in an aircraft that was being tossed about by updraughts and downdraughts or in evasive action.

This was the first operation on which the great Oswald

Boelke led a formation of his newly formed squadron: one flight of six aircraft, in which Manfred was flying as his No. 2. German aircraft seldom ventured behind the Allied front line; the British and French had to cross into enemy territory to get at them. This morning, the Albatroses intercepted eight BE2cs of 12 Squadron, escorted by six FE2bs of 11 Squadron, far beyond the German line and cut them off.

In his autobiography, Richthofen describes the FEs as 'seven great two-seat bomber aeroplanes'; an absurd exaggeration of size and a crass misidentification of type. It was the BE2s that were carrying bombs: 20 lb Coopers stowed in the cockpit and dropped over the side by hand. Also, the positioning of the two types must have made it clear which were the bombers and which their escort.

FEs (known to the RFC as 'Fees') had been operating in France since the spring, so even the enemy airmen who had not yet seen one should have known all about them. The German intelligence branch was presumably at fault for not having propagated the information. The BE2s, smaller and more frail than their escorts, must have been a familiar sight: the RFC had been flying them since 1914. Their best speed was 72 mph and it took them 45 minutes to reach 10,000 ft.

They were a tractor type, originally unarmed, but the BE2c carried a Lewis gun in the front cockpit. A spigot projected from its underside and there were holes around the rim of the front cockpit in which this could be located. The observer had to heft the bulky 27 lb gun and its ammunition pan from hole to hole, according to whether he wanted to fire to the front, abeam, or astern. That was hard work, swaddled in several layers of bulky garments, panting in the thin air above 5,000 ft, shifting the weapon without dropping it overboard while the light little aeroplane dipped and swayed. For firing astern, a hinged swinging arm was fixed abaft the cockpit and life became acutely disconcerting for the pilot, with a hot blast and crackle as bullets whizzed over his head; and worse when tracer ammunition scintillated past his eyes.

Richthofen attacked an FE and they exchanged shots.

'I tried to get behind him, but this fellow was no beginner, for he knew very well that the moment I succeeded in getting behind him his last hour would be sounded. My Englishman twisted and turned, crossing my line of fire.'

Eventually he managed to get onto the FE's tail and shoot pieces off its engine, which stopped. It hit the ground near another German squadron's airfield.

He landed near it, an act that is often attributed to chivalrous concern for its occupants. It was nothing of the sort. Richthofen wrote, 'I was so excited that I could not resist coming down. I landed near the Englishmen and jumped out of my aeroplane.' He had killed the observer, Lieutenant T. Rees, and mortally wounded the pilot, 2nd Lieutenant L.B.F. Morris, who died while being taken to the nearest hospital. However, he made a handsome gesture in recognition of the fight that his opponents had put up: he visited their graves and had headstones erected, 'to the memory of my honourably fallen enemies'.

This was a typical fight between an unwieldy two-seater armed with a light machine-gun and a nimble single-seater equipped with two heavy machine-guns.

The other pilots whom Boelke had led were also novices. On returning to base Manfred learned that 'Every one of the beginners had gained his first aerial victory'.

To commemorate his, he ordered from a Berlin jeweller a silver cup 5 cm high and 3 cm wide, engraved '1 Vickers 2 (meaning two-seater) 17.9.16'.

The FE was, in fact, designed by de Havilland and manufactured at the Royal Aircraft Factory. The only Vickers aircraft in France was the FB5 Fighting Biplane, called the Gunbus by the RFC, which had been at the Western Front since 5 February 1915. Like the FE, it was a pusher and the observer occupied the front cockpit with a Lewis gun on a low mounting, limited to fire through a 180 degree arc forward. On the FE2 a distinctive cylindrical fuel tank projected horizontally behind the pilot's head. Both types had lattice booms instead of an enclosed fuselage, to which the tail unit was attached; but the shape of the tail fin and tail planes differed. There was no excuse for misidentifying either aeroplane.

The Royal Flying Corps's opinion of the German Military Aviation Service *(Luftstreitkräfte)* at that time would not have pleased the enemy. In 1916 a manual, *Fighting in the Air*, was issued, written by one of the most experienced and successful British pilots, Major L.W.B. Rees, commanding 32 Squadron. His use of capitals is eccentric but most of what he says must have been of great value. He begins:

> 'These notes are based on experience of last year, so that it is impossible to lay down any hard and fast rules, as the conditions alter so fast. The deductions are based on the experience of many RFC officers, to whom I am greatly indebted.'

The document was reprinted and brought up to date, as far as possible, in July 1916. Under the heading 'Comparison of Pilots', it says:

> 'The British Pilot always likes the idea of fighting and is self-reliant. He is a quick thinker compared with the enemy, so that he has the advantage in manoeuvre. He fights for the sport of the affair, if for no other reason. After the first engagement he gains great confidence from the Parthian tactics of the enemy. Very wisely, he is not hampered by strict rules, and as a rule is allowed to conduct his own affairs.
>
> 'The Enemy Pilot, on the other hand, is of a gregarious nature from long national training, and often seems to be bound by strict rules, which cramp his style to a great extent. The Enemy Pilots are often uneducated men, being looked on simply as drivers of the machine, while the Gunner or Observer is considered a higher grade than the Pilot.
>
> 'This last gives the advantage to us, as, whereas our Pilot acts from a sense of *Noblesse Oblige*, the Enemy, when in a tight corner, often fail to seize and press an advantage.
>
> 'We noticed that when there were two officers in the Enemy

machine, they always attacked, but in many other cases the attack was not pressed home. Untrained Enemy Pilots might also account for this.'

The last paragraph does not merely cast doubt on the quality of German air crew training. There is also a manifestation of old-fashioned class prejudice in imputing lack of determination or courage to NCO air crews; and a hint of the patronising, amusedly contemptuous British attitude to other nations not unknown today: 'One can't expect a foreigner to know better.'

About German gregariousness, Major Rees seems to have been right. The designers of *Luftwaffe* bombers more than twenty years later accommodated the whole crew in the smallest possible space, to create an atmosphere of mutual support. In only one German type were the rear and dorsal gunners in lonely separation from the rest aboard, as in the British Whitley, Wellington, Stirling, Halifax, Lancaster and Sunderland; and the American B-17, B-24 and B-29. The sole exception was the Focke-Wulf Kondor, which had an upper turret towards the tail; but it flew exclusively on maritime reconnaissance far from land, where it was seldom attacked; and only 276 were produced, compared with 6,086 Heinkel 111s, 5,709 Junkers 87s and 14,980 Ju88s.

Under the heading 'Comparison of Duties', Rees explains that both the opposing air forces

'divide the machines into two classes. Reconnaissance, by which I mean those that do reconnaissance proper, wireless, photography and bombing, and Fighting, used for fighting only. The Fighting machines are used for Patrol, and escorting the Wireless, Bombing, and Reconnaissance machines.

'The enemy uses his machines differently to (*sic*) ourselves. His Reconnaissance machines come over our side of the Line only at comparatively long intervals, they seldom come over far, and they travel at great heights. Sometimes fast machines come over singly, and sometimes the slower machines come over in Flights of 6–8.

'The fast machines are so fast that only our fastest machines can catch them. If fired on they immediately dive for their Lines, or for the nearest Anti-Aircraft Battery or Machine Guns. As every village near the Lines has its machine-gun, it means that the machine can dive almost anywhere so as to get a covering fire from the ground.'

He judges enemy tactics fairly. 'These machines very seldom turn and fight, very rightly going straight back with their information.'

On the other hand he is hard on their aircraft handling. 'Many of the Enemy Pilots are heavy handed, so that the machine turns over on landing, if the ground is at all rough.' This was not uncommon when pilots of any nationality and calibre had to forced-land on rough ground: aeroplanes were so light, they would tip over even with the best of pilots at the controls.

One of his most cogent observations is,

'Our Reconnaissance machines, on the other hand, are continually over the enemy lines, so that our Fighters have to go for miles to get a fight. This affects the tactics, in so much as the Enemy can risk getting hit on the engine or through the tank, knowing that he will suffer no more than an ordinary forced-landing. Our Fighters have to be more careful, as hits on the engine usually mean that the crew of the machine must be taken prisoner.'

Later, he writes,

'There should be no long range shooting, and if we can manage to disable the Enemy quickly, there will be no need to go out of action in the middle of an engagement while the drum is being changed.'

He describes German methods of attack. Single-seaters

'. . . try to creep up behind their targets unseen. If seen or fired on they dive immediately and come up again after a short while. They do not as a rule accept a set battle.

'The second type is a machine a little larger. It also is a monoplane, but carries either a Pilot alone or a Pilot with a Gunner. They fly in flights of four or six, and travel at a great altitude. When they attack a machine they dive at it firing the whole time, one after the other. They do not stop to re-load, but go straight down, even if they are not fired on. They do not usually return to the attack. They fire straight ahead and straight up, but do not usually fire astern.'

He repeats and emphasises that the duty of a fighter is to put the enemy's aircraft out of action, and that most of the fighting is done on the German side of the lines.

'It is not sufficient to make a machine land, as machines are comparatively easy to obtain. Every effort should be made to disable the Enemy Pilot, as this nearly always ensures the destruction of the machine as well, even if dual control is fitted. In any case it prevents the Enemy using his armament effectively and stops the machine manoeuvering.

'If the pilot is taken as the target, the shots which miss the target will hit the Observer and engine, or may cause damage to the rigging.

'To be of real use the Pilot of a fighter must be extremely keen sighted. I believe one can intimidate the average Enemy Pilot more by showing that he has been seen than by doing anything else.

'When one sees a machine one is apt to think that hits anywhere will be effective. One is trained to imagine that a small thing, such as a frayed cable, is certain to cause a wreck. Yet machines go up every day and return absolutely under control, but having dozens or even hundreds of holes in different places. It should be remembered that after being over the Enemy's Lines, machines should be brought back with the greatest care. Machines are sometimes wrecked over their own aerodromes because a thoughtless Pilot does a steep spiral, perhaps not knowing that his main spars have been pierced.'

This was a forewarning of the danger in doing a victory roll over the airfield on return from a successful fight, which killed over-exuberant pilots and wrecked Hurricanes and Spitfires in the early months of the Second World War. It was quickly forbidden. Rees continues:

'The only useful target to really attack is the Pilot himself. This target is very small, being of a size about 2' by 1'6" by 1'6", and even then shots which hit this target are not certain of putting the Pilot immediately out of action. Therefore one must concentrate one's attention and one's shooting on this small target, the Pilot, till one has obtained one's object.

'If we attack a machine from directly in front or in the rear the engine may cover the Pilot's body or vice versa. This is the minimum target which the machine can present, and any shots hitting the target do damage, but there is a lot of room round the target in which shots which do not actually strike do no damage.

'Now, if we imagine a machine being attacked from the side, or straight from above or below. The target which we must aim for still remains the same small one, but now the rounds, which before were non-effective, will hit the engine and Observer and become effective.

'This leads one to suggest that the way to attack is straight at an Enemy from above, below or from the side, keeping one's own machine end on to him.

'It is very hard, when looking at a machine in the air, to know where the Pilot is sitting. This may sound incorrect, but if approaching from below one sees only the bottom of the Enemy's fusilage [sic], and as the machine is unfamiliar the exact spot we want to hit is hidden.

'With a small Fighter we should close as soon as possible, keeping end on to the Enemy, so that he will have no chance of setting any sights he may have. We are then never at a disadvantage, and we have the advantage of being the attacker. A machine coming at one quickly always makes one a little nervous, especially if one does not know the Pilot.'

Presumably he means that some German aces such as Boelke and Immelmann were recognisable by their aircraft or flying style; Richthofen had not yet joined their ranks.

Major Rees deals briefly with the difficulty of estimating relative speeds. If two aeroplanes are circling at the same speed,

'the relative motion is apparently nil, but the actual relative motion at the moment of firing is practically the same as though each machine were flying straight. The enemy apparently sits on the gun sights without motion, but the maximum allowance for speed must be made.

'Then, again, as both machines are banking over, it will be very hard to estimate if there should be an allowance, because the gun is apparently elevated.'

About the range at which to shoot:

'The range at which fighting takes place may vary from 400 yards to 4 yards. It is very hard to approach a machine to 100 yards without being seen. Hundreds of rounds are fired every day at machines at ranges estimated at 50 yards or less without doing any damage. At 200 yards one may expect to get hits, and I have taken that as the normal fighting range. I do not think that there has been a single instance in which machines have been brought down at ranges over 400 yards. Thus we see that it is useless waste of ammunition to fire at long ranges, and that one should try and close to within 50 yards in order to do any damage.'

On the matter of tracer bullets: when they came in

. . . 'it was thought that they would make close fighting impossible. They have not made the difference that one would expect. One reason is that it is very hard to estimate the range in the air, just as at sea. The tracers burn for a comparatively short time, so that they go out before hitting the target. This means that the bullet apparently hits, but really falls away from the target.'

Under 'Usual Enemy Tactics' he says that hitherto fighters have tried to approach from behind and, if seen, dived then climbed for a belly shot and continued to dive and climb repeatedly. These were predominantly the Fokker E1. However, with the entry of the Albatros: 'Sometimes now they dive under one, and then climb quickly, so that when next seen they are above and behind one's machine. To prevent this, hustle the Enemy to prevent him coming up again.'

The tactics laid down for RFC pilots flying single-seat fighters and the observer-gunners in two-seaters were the same as those that the enemy had worked out for themselves, except that their Spandau and Parabellum machine-guns were of a heavier calibre than the Lewis and effective at longer ranges.

It was at this point in the development of air fighting that Manfred arrived on the scene.

CHAPTER 3

HEREDITY AND FORMATIVE YEARS

anfred's family were typical *Junker*, aristocratic Prussian landowners; and, though not, in the words of Sir William Gilbert, 'the very model of a modern major-general', he was undeniably the very essence of a German regular officer. Obedience, discipline and duty were inculcated in him from childhood and set his standards of conduct for life.

Like most of his kind, he was an ardent hunter of game, from partridge to bison. It was this passion for hunting that has caused many critics to condemn him as a cold and calculating, even psychopathic, killer who took human life as unfeelingly as he shot birds and animals. This was jumping to a wrong conclusion, because a good game shot prides himself on making clean kills that do not cause suffering. There are others who attribute his enjoyment of air fighting to a love of sport, as Major Rees did about British pilots, but that is not true either. War has never been a sport, it is a grim business in which men – and women – set out to destroy the enemy from patriotism or political dedication. A clean kill is not given a thought. One puts the enemy out of action by whatever means one can: death, swift or slow, or wounds that make him unfit to fight permanently or temporarily.

In battles between aircraft, before the present era when homing missiles blow pilots and crews to smithereens in a split second, men often burned slowly to death, died in agony from mutilation, fell thousands of feet without a parachute, or suffered some other tormented end. The victors who inflicted such horrors on them were decent people of various nationalities, including British, who would never have been deliberately cruel to anyone.

A passage in Manfred's autobiography vindicates him of accusations that he was insensitive and bloodthirsty: 'I have long since had nightmares of the first Englishman I saw plummeting down'. The context makes it clear that he was referring to the first enemy aircraft he shot down. His heredity and upbringing to be a professional soldier had not blunted his humanity.

His parents were *Major* Albrecht *Freiherr* and Kunigunde *Freifrau* von Richthofen. His father was serving in the dragoons, the First *Leibkürassieren* Regiment, at Breslau, capital city of Silesia, with 600,000 inhabitants (now Wroclaw, in Poland), when their first three children were born: a daughter, Ilse, in 1890, Manfred on 2 May 1892, Lothar in 1894. The Major, after rescuing one of his troopers who had fallen into the wintry river Oder, became deaf in one ear and was invalided out of the Army. The family moved to a small town thirty miles south-west, Schweidnitz (now Swidniza), population then 30,000, where the youngest child, Karl Bolko, was born in 1903.

From early boyhood the eldest son was adventurous. One of his frequent feats was to climb the tallest apple tree in the orchard, then, instead of clambering down the same way, swing from branch to branch.

At the age of eleven he entered the Cadet Academy at Wahlstatt. He had not chosen a military life. 'I was not particularly eager to become a cadet, but my father wished it and I was not consulted'. He was so unhappy there that he warned Lothar not to be coerced into following him.

He did not do well academically, because he worked only enough to pass his examinations, but excelled at gymnastics and marksmanship. He was particularly expert on the parallel bars and could also perform the difficult standing forward somersault.

A year later, during a holiday between semesters, he was given an airgun. To demonstrate his prowess, he shot three ducks and proudly brought his mother a feather from each as trophies. Presumably he was inspired to emulate his father, who had adorned the house with some 400 animal heads and stuffed birds as witness to his skill with rifle and gun. Asked where he had found the ducks, he confessed that they were swimming in the garden pond. Evidently German boys were not taught that it is the worst of bad sportsmanship to shoot a sitting (or floating) bird. His mother scolded him, but his grandmother intervened with the plea that his honesty in admitting to the slaughter of what were regarded as pets outweighed the misdeed. He glued the feathers to a sheet of brown paper and they became the first of his many trophies, which culminated in his collection of cups and aeroplane parts commemorating his aerial victories.

While at the cadet school, to test his nerve he climbed the highest church tower in the town and tied a handkerchief to the lightning conductor as proof of his daring. There is an aberration in his recollection of the escapade. In his book *My Life In The War*, he wrote amusedly, 'Ten years later I visited my youngest brother, Bolko, at Wahlstatt and I saw my handkerchief still tied high in the air.' Later he contradicts this with equal amusement:

> 'My youngest brother, Bolko, wrote a long letter of complaint to the family. He is a cadet at Wahlstatt and says that I portrayed the teachers so badly in my book, he is having so much unpleasantness that he cannot bear it any longer. He asks that I submit the manuscript of my next book for his approval. I think he demands a lot, besides accusing me of lies. I related how I once climbed the church tower at Wahlstatt and hung a handkerchief there. He has established that it no longer hangs there and because of that I could hardly have told the truth. I think he is asking too much of a handkerchief to adorn a church tower for fifteen years.'

The display of courage would have earned the admiration of Douglas Bader and others at the Royal Air Force College, Cranwell, in the late nineteen-twenties. The bravest cadets, Douglas among them, performed a lethally dangerous trick for which they would have been dismissed the Service if found out. Flying solo in the rear cockpit of a dual control Avro 504K (top speed 95 mph), a trainer version of a 1914–1918 reconnaissance type, these young pilots undid their seat straps and parachute harness, climbed out of the cockpit, straddled the fuselage and moved forward to tie a handkerchief to the control column in the front cockpit; then returned, refastened all the straps, landed and removed the incriminating evidence before an officer spotted it. Of such intrepid high spirits have fighter pilots always been made; but in the later 1930s high speeds made such defiance of danger impossible. A potentially fatal climb by a fourteen-year-old was a fair equivalent.

In 1909 Manfred was admitted to the Senior Cadet Academy at Lichterfelde. He had been riding since boyhood, and here he competed in equestrian events. On passing out from there he became an Officer Candidate, applied to join No. 1 *Ulan* (Lancer) Regiment *Kaiser* Alexander III of Russia and was accepted. (Alexander III was the *Kaiser*'s uncle and father of *Czar* Nikolai II, Russia's ruler.) The next stage of his training was a short course at the Berlin War Academy, after which he was commissioned. He records, 'Finally I received my epaulettes and was so proud to be addressed as *Herr Leutnant*.'

At twenty years old he was a personable young man with a ready smile that graces even photographs taken at times of intense battle stress. He was short, like many great fighter pilots in both world wars, slim and blue-eyed; again, blue or grey eyes are a common feature among the best fighter pilots and are said to denote excellent sight. He appears to have been more than a trifle bandy-legged (not quite in the couldn't-stop-a-pig category) which is not surprising in anyone who has ridden a lot from an early age.

He took to regimental life with zest. His father gave him 'a beautiful mare named Santuzza'. He found that she was a fine jumper, so entered her for a jumping race – whether over fences or hurdles he does not say. On the eve of the event she fell while practising, injured a shoulder and he cracked a collarbone.

He seems to have been a bold and reckless horseman rather than a polished one. At Breslau in 1912 he competed in what he calls 'the Olympiad'. The word was already in common use as a synonym for 'Olympic Games', although its original meaning was the four-year period between each Games. His use of the term is puzzling. This was an Olympic year, but the Games were held in Stockholm.

He rode 'a very beautiful chestnut bay horse in the distance race'. Approaching the last jump, he saw a crowd of spectators there and deduced that it must be the most difficult one. People were gesturing at him to reduce speed, but he ignored them. He galloped up a slope at the top of which was the obstacle. Felix flew over it, then 'mount and rider disappeared, head over heels, into the Weistritz River', just beyond. But he kept his sense of humour: 'At the weigh-in, people were amazed that I had not lost the usual kilogramme, but, rather, was five kilos heavier. Thank God no-one saw I was soaking wet.'

His last win came on a mare named *Blume* (Flower) in the 1913 Kaiser Prize Race. He was the only rider to have a clear round, despite an accident.

> 'I galloped over the heather and then suddenly landed on my head. The mare had put her foot in a rabbit hole and in the fall I broke my collarbone. I remounted and rode another seventy kilometres with the injury, but finished in good time without a fault'.

With seventy kilometres remaining after he had already completed part of the course, the mind reels at trying to guess how long the whole ride must have been.

He makes no mention of girl friends or mess life. The impression one has is that he was too private a person to record any romantic relationships or sexual liaisons and would, anyway, have regarded it as ill-bred and ungallant to mention such intimate matters. However, it is unlikely that in his circumstances a well-off, good-looking, lively young sportsman would choose to be celibate; especially in an era when uniforms and the dashing reputation of cavalrymen exerted a highly charged sexual attraction.

A well-known RAF fighter pilot, interviewed during the fiftieth anniversary of the Battle of Britain, said frankly 'At nineteen, all you think about is beer and women'. Dashing young airmen had superseded the cavalry, who now clanked about in armoured vehicles.

This echoes what a famous World War I Canadian fighter pilot wrote, unambiguously, in his memoirs;' On some evenings we would go into the local town to drink and dine, and do more interesting things.'

There was, in fact, a sweetheart in Manfred's life, but, as will be seen, the association was, to use a contemporary term, strictly honourable.

Carousing is another matter and young officers of all nations have traditionally been subjected to it by their seniors as a rite of passage, for 'a gentleman must learn to hold his liquor'. Manfred mentions the odd occasion when champagne flowed, but only to celebrate a special event or ameliorate tedium, not in the daily life of an officers' mess. Perhaps he had another characteristic in common with Douglas Bader, who did not drink; he had tasted beer, sherry and whisky but disliked them. A brilliant cricketer, rugger player and boxer, he eschewed alcohol anyway, because it reduced fitness; and his natural high spirits were enough to make him enjoy a party or the company of his comrades without artificial stimulus.

Chapter 4

War is Declared

I n his *History of the First World War*, B.H. Liddell Hart
wrote, 'Fifty years were spent in the making of Europe
explosive. Five days were enough to detonate it'. This is
relevant to Manfred von Richthofen's career because, unless
the causes are explained, one might wonder why his regiment,
named after a past *Czar*, should suddenly find itself fighting
Russia.

The explanation is that on 27 June 1914 Archduke Franz
Ferdinand, heir to the ruler of Austria-Hungary, and his
wife, were assassinated on a visit to Serbia. On 28 July
Austria-Hungary declared war on Serbia. Two major opposing
alliances existed in Europe: one between Germany, Austria-
Hungary and Italy, the other comprising Great Britain, Russia
and France. Russia, an ally of Serbia, mobilised its armed
forces. Germany, supporting Austria-Hungary, declared war
on Russia on 1 August and on France two days later. On
the following morning Germany invaded Belgium. Thereupon
Britain declared war on Germany. Italy deserted its alliance
with Germany and Austria-Hungary and declared itself neutral;
but came in on the side of Britain, France and Russia on 24
May 1915.

In expectation of war, Manfred's regiment had been moved to
a small town, Ostrovo, near the Russian frontier. In his book he
makes a rare mention of what he calls 'celebration' in the mess
on the evening of 1 August, which is a euphemism for tipsy
revelry. The regiment had been brought to readiness several times
recently and each alarm had been a false one. 'We had convinced
ourselves that war was out of the question.' They had to change

their minds when their party was interrupted by the District Magistrate, who brought urgent news. Germany had closed the bridges in Silesia across the river that formed the frontier with Russia and guards had been posted at every one.

On 2 August Manfred wrote home to say, 'These, written in haste, might be my last lines. My heartfelt greetings to you all.

If we do not see each other again, accept my most sincere thanks for everything you have done for me. I have no debts and am taking money with me. I embrace each of you. Your grateful and obedient son and brother.'

At midnight he led a mounted patrol over a bridge across the river Prosna, expecting an immediate encounter with the enemy. Early next morning the *Ulans* entered Kieltze, a village whose church tower attracted his attention. He posted a lookout there, then locked the parish priest in it, had the belfry ladder removed so that he could not ring an alarm, and warned him that 'if the slightest hostile behaviour were noticed among the population', he would be hanged.

Although the German Imperial Military Air Service comprised thirty-three Field Service Units, each of six aeroplanes, and ten Home Defence Units of four, none had yet been sent to this sector of the Eastern Front, so the calvary had to do all the reconnaissance.

The Russians also relied on cavalry here at first. When they did begin aerial reconnaissance, one of their pilots was unique: Princess Shakovskaya, the only woman flyer in any air force. Her photograph appeared in a December issue of *The War Illustrated*, leaning with elegant nonchalance against her aeroplane.

Manfred had to send a despatch rider to his Headquarters every evening with a report. As these did not rejoin him, their number steadily diminished.

On the fifth night, by when all but two of his troopers had gone, the sentry on the church tower woke him and announced, 'The Cossacks are here.'

It was drizzly and starless. The three Germans led their horses away. It was so dark that fifteen metres sufficed for a safe distance. Manfred's weapons were the sabre and pistol. He borrowed a carbine from one of his men and returned alone to reconnoitre. He had released the priest that day and now, by the light of lanterns that some of the enemy carried, saw him among them, whose numbers he put at twenty to thirty.

He rejoined his men and they took their horses to a wood, from where, at dawn, they saw the Cossacks leave. That evening, he sent a sixth man off with a despatch and on the seventh returned to the regiment with the one remaining trooper. There was astonishment at his appearance, as a rumour had spread that they had both been killed.

That was the last he saw of the Russian Front for the time being. The regiment entrained next day for France and arrived at a destination north of Metz twenty-four hours later. Festivity relieved the monotony and discomfort; 'We had to take provisions for a long journey. Naturally, we did not lack something to drink.'

First they rode north for some thirty kilometres towards Luxembourg with patrols scouting ahead, but had no sight of the enemy. After crossing into Luxembourg they headed north-west to Arlon, in Belgium, another forty kilometres. Everywhere, 'right and left, on every street, in front of us and behind us' marched soldiers of every arm.

On 21 August, at Etalle, seventeen kilometres west of Arlon, he led a fifteen-strong patrol to learn the strength of the enemy who were in a vast forest. They came upon a forester's uninhabited cottage and after they had ridden past it a shot came from one of its windows; a trooper was wounded in the hand and his horse in its belly. The patrol surrounded the house and found half a dozen youths inside, but no gun.

'My anger was high, but I had never killed anyone in my life and the moment was extremely unpleasant.'

The youths bolted through the back door. The troopers later found a gun, so Manfred ordered the cottage to be burned down. He had been set a precedent; nine days before, his cousin Wolfram was killed by a *franc tireur* at a village in this vicinity and every house that might hide these partisans had been put to the torch.

Now from the edge of the trees he saw a troop of French dragoons. When they disappeared he set out to find them. On his right was a wall of rock, on his left a stream; fifty metres ahead, a meadow; then the forest's edge. He was looking through his field glasses when rifle fire broke out from the forest, aimed at his patrol. He estimated that over 200 men were hiding there. It would be useless to take cover, so he ordered a retreat. Against his orders, his men had bunched together, so only he and four troopers escaped; but a couple more came back on foot, their mounts shot from under them. 'This baptism of fire was not as much fun as I had thought.'

His comment on his men's demeanour when they realised they were much outnumbered, was, 'Naturally, no one thought of anything but attacking. It is in the blood of every German to rush out to meet the enemy, especially enemy cavalry'. He was to learn that this was not unique to his countrymen; in the air as well as on a charger.

Reluctance to kill civilians did not trouble him for long. After some more hard fighting against an outnumbering French force, he and a *Leutnant* Loen were sent out with a patrol but did not find the enemy until the evening. They decided to spend the night in a nearby monastery, where for the first time in three days their horses could be unsaddled.

'The monks were very friendly and gave us as much to eat and drink as we wanted. It should be noted that three days later we hanged several of our hosts from lampposts because they could not resist the urge to take part in the war.'

The Gestapo or SS could not have been more brutal twenty-five years later.

It is anomalous that the young man who had threatened to hang a Russian village priest and now carried out this barbaric act in Belgium – an outrage against hospitality, too – could also be troubled by nightmares about men whom he was to kill in air combat in the near future. The explanation is presumably that he despised civilians but respected all military men and had a fraternal feeling for pilots and air crew of any nationality. Snobbery also came into it, one suspects: enemy airmen were mostly officers and therefore accepted as gentlemen, even if only temporary ones; whereas most clergy were of humble origin. It is doubtful that he would have executed a bishop or cardinal, whose prospects of high rank depended largely on superior education, which meant enough money to pay for it and in turn suggested the upper class.

On 24 September he wrote home to announce that he had been awarded the Iron Cross, Second Class. Referring to a letter he had received from his mother, he conveys the hardships and dangers of campaigning. She expressed surprise that he was saving 'so much money' – it cannot have amounted to a great sum in eight weeks. He explained that after the war he would have to re-equip himself completely. He told her that all the gear he took with him – which must have been an exaggeration – had been lost, burned or torn; even his saddle and the rest of his tack. So what was he wearing and was he riding bare-back, hanging on to his horse's mane?

On the same day, his father, who had applied to return to the Active List, was appointed administrator of the Silesian Reserve Hospital in Kattowitz.

Manfred's expectation in any war in which he might fight had been of exciting cavalry charges that would be as enjoyable as riding in cross-country events; and, of course, swiftly over with his regiment victorious. The squalid and muddled reality of scruffy billets in a primitive village, no sight of the enemy for several days and then a hasty withdrawal under cover of darkness at the threat of a losing fight against a greatly superior force, held no glamour and gave no satisfaction. Next, tedious patrols and the disastrous encounter in a Belgian forest against heavy odds that again ended in a retreat, were another disillusion even though he earned a medal. Soon came further frustration. To understand why he found himself immobilised, unhorsed and so miserable that his thoughts turned away from cavalry engagements to air battles, German offensive strategy and France's defence must be examined. How did he find himself in what he describes in a letter home as 'a bomb-proof, heated dugout, for weeks on end'?

The reason lies in the nature of France's 150-mile frontier, whose south-east end abuts on Switzerland. A stretch of flat country, the Belfort Gap, runs seventy miles along the Vosges Mountains. Behind and prolonging this natural rampart ran an almost continuous fortress system based on Epinal, Toul and Verdun. Twenty miles beyond Verdun lie the Luxembourg and Belgian frontiers and the difficult Ardennes country. The only feasible gap in the barrier was the Trouée de Chambres between Epinal and Toul. This was left open originally as a strategic trap in which an invader could be caught and defeated by a counterattack. The other possible line of approach was by Belfort and Verdun. As has been related, the Germans went round these formidable obstacles, through Luxembourg and Belgium, then were stopped at Verdun, the ancient gate of the West through which Attila had led the Hun hordes when they tried to conquer Gaul in the Fifth Century. The Order of the Day was *On ne passe pas*, 'They shall not pass'; and they never did. In 1916 a battle that lasted from 21 February to 15 December would be fought and end when the Germans abandoned their efforts to overcome this massive fortification. Meanwhile, in September 1914, the stalemate had begun.

There was no use for the cavalry in this siege, apart from an occasional reconnaissance, so we find Manfred glumly writing to his mother, 'My activity before Verdun was boring. At first I was in the trenches at a place where nothing happened'. Then he became an assistant adjutant and expected to be actively involved in the fighting again, but instead was demoted and spent his days beside a telephone in the deep dugout. He was allowed no nearer to the front line than 1,500 metres.

He alleviated his dull existence in the way that most appealed to him. Riding in the forest of La Chaussée, he saw wild pig and at once began to make a plan for shooting them at night. 'Beautiful snowy moonlight came to my aid. With the help of my orderly I built a shelter seat and waited. I spent many nights in trees, feeling as cold as an icicle'. Eventually he was rewarded. 'One sow swam across the lake every night, broke into a potato field at a certain place, then swam back.' He waited at the lakeside and shot her while she swam. 'She would have drowned if I had not at the last moment seized her by the leg'. Another day, when riding with his orderly along a narrow path, several wild pigs crossed in front of them. 'I jumped down from my horse, grabbed my orderly's carbine and ran several hundred paces ahead'. A boar emerged from the trees. 'I had never seen a boar before and was amazed at how gigantic this fellow looked. Now he hangs as a trophy here in my room: he is a beautiful memory'.

On a day in October when he led one of the infrequent calvalry patrols, he might have preferred to stay in shelter, however tedious and demeaning to his pride. He had just halted his men and dismounted when 'a grenade exploded on the saddle of my horse', and killed it and three others. The enemy was unpleasantly close at hand. 'My saddle and the contents of the saddlebag were ripped to pieces.' Evidently he had already replaced the tack that he had reported as having been destroyed some weeks previously. 'A fragment ripped through my cloak but did not touch me'. It seems he had new clothes, too. Apparent inconsistencies and contradictions in his letters and his book make his memory suspect: total recall, he manifestly did not have.

He was billeted with another officer in a village where most of the houses had been burned down, and spent alternate days in the trenches, 'like the infantry'. His only remaining horse was sick, so he had little chance to ride. In a letter on 2 November, he complains that for weeks there has been no more than 50 metres advance at Verdun. 'I would like very much to have won the Iron Cross, First Class, but there is no opportunity.' He adds, with wry levity, 'I must go to Verdun dressed in French uniform and blow up a gun emplacement'.

In mid-January 1915 his spirits rose and he wrote to his mother, 'I have already let you know I am Assistant Adjutant of the 18th Infantry Brigade'. This offered more activity than with his regiment. 'I am quite satisfied with my post'. One night, 'we of Grenadier Regiment 7 took a trench from the French'. Two nights later the enemy tried to recapture it but 'were splendidly crushed'.

He was devoted to his family and his thoughts were constantly with them. He wrote frequently to his mother, and to his father about his boar hunting. In his letters he often mentioned Lothar and sent love to 11-year-old Bolko, who was at the Wahlstatt Cadet Academy. In March he informed both parents that he had been host at a shoot with five other officers and thirty beaters, who drove eight pigs out of cover, 'but we all missed. The hunt lasted from eight in the morning until seven in the evening. In three days we will have another try, and in ten days, with a full moon, I am expecting confidently to bag a wild boar'.

His tally of pigs killed still stood at three when, for him, the crunch came. He had been home for three weeks' leave and, on his return, was expecting to take part in a minor offensive, but instead was given more office work.

Bitterly frustrated and aching to go into action again, he applied for a transfer to the Aviation Service, which was approved in May. An unexpressed desire to fly had been growing for a long time; so now, 'My greatest wish was fulfilled.'

Chapter 5

The Reconnaissance Observer and Pilot

At the outbreak of war, none of the officers and men in the armies involved knew much about their flying services, or about aeroplanes at all. Few recognised military aircraft markings, which were, anyway, difficult to see from the ground. The British and French had roundels – known also as cockades, an anglicisation of the French *cocarde* – on wings and fuselage: concentric circles with a 'bullseye' centre. The British bullseye was red, with a white ring round it and a dark blue outer ring. The tail fin bore vertical stripes of red, white and blue. The French wore a lighter blue bullseye with concentric white and red circles. On the tail fin were blue, white and red vertical stripes in reverse order to the British. The Germans had a Maltese Cross in black on a square white background, on wings, fuselage and tail fin. In 1916, the white square was replaced by a white edge around the cross.

All air forces were at first regarded as an extension of the cavalry and were therefore used only for reconnaissance, on which they often flew low enough to come under rifle fire. They were frequently the targets of their own ground troops; many were hit and some airmen were wounded.

Manfred confessed, 'I had not the slightest idea what our flyers did. I considered every flyer a deceiver, for I could not tell whether he were friend or foe. I had no idea that German machines bore crosses and the enemy's had cockades.'

To fly was the obvious choice for a cavalryman now that his arm was becoming obsolete and must soon be completely so, in a war that he and his comrades had taken for granted would be over in a couple of months. To transfer to the infantry would be beneath his dignity, as well as abhorrent on account of the sordid living conditions in the front line. To join the artillery would entail months of study and he had no taste for scholarship. Nor would he be content to shoot at targets miles away where he could not witness the havoc his shells were wreaking. The air offered the same sense of liberty, individual initiative and expertise that cavalry officers enjoyed and called for the same dashing spirit. He would also be able to watch the effect of his action.

He had no thought of being a pilot. Not only because being an observer would put him in command of the aircraft but also because he was in a hurry to return to action and the course lasted only three weeks, whereas pilot training took at least three months.

'I was anxious to get into the air at the front as soon as possible and began to fear that I might be too late and the war would be over before I could get going.'

On 31 May, thirty observer candidates assembled at No. 7 Aircrew Replacement Unit in Cologne, Manfred among them. On the night before his first flight he went to bed early 'to be fresh for the great moment next morning'. At 7 am he was at the airfield. The trainer aeroplane was a tractor type, so, sitting in the front cockpit, he felt the full force of the propeller wash when the engine started.

'The blast of wind from the propeller disturbed me greatly. It was impossible to make myself heard by the pilot. When I took out a piece of paper [to write a message to him], it disappeared. My flying helmet slipped off, my muffler unloosened and my tunic was not securely buttoned. I was miserable. Before I knew what was happening, the pilot opened the engine up to full speed and we began to roll faster and faster. I hung on, then the shaking stopped and we were in the air, the ground slipping away beneath us.'

He had been given the course to fly, but after his pilot had turned a few times, 'I had no idea where I was.' Cautiously, he peered over the side. 'The people looked tiny and the houses like children's toys. It was a glorious feeling to sail over everything.' He felt reluctant to return to the ground.

On 10 June he moved on to No. 6 Aircrew Replacement Unit at Grossenhain. Eleven days later, the first of his batch to pass the course, he was posted to No. 69 Field Reconnaissance Unit, on the Eastern Front. They lived in tents near a village where most of the houses had been burned down and, Manfred told his mother in a letter, those still standing were verminous.

It is difficult to understand how his pilot, *Leutnant* Georg Zeumer, had been passed fit for any form of military service, since he was dying from tuberculosis. Knowing that he had not long to live, he flew recklessly, which Manfred admired and enjoyed.

Their aircraft was the Albatros BII, designed by Ernst Heinkel, whose name was to become loathsomely familiar to the inhabitants of bombed British towns in the next World War. It had a water-cooled, in-line 100 hp Mercedes engine. The radiator was mounted above the cylinder block with the exhaust manifold on top and the exhaust pipe rose over the upper wing. The BII was a fraction over 25 ft long, had a maximum speed of 65 mph, a service ceiling of 9,840 ft and four hours' endurance. No fixed defensive armament was carried, but during the early months of the war the observer, in the front cockpit, usually armed himself with a carbine or rifle.

Five weeks after he began his operational tour, Manfred wrote, 'I fly over the enemy almost daily and bring back reports. I reported the retreat of the Russians three days ago. It is more fun than being an Assistant Adjutant'.

Zeumer had been returned to the Western Front and Manfred began flying with a recently qualified pilot who was very much a man after his own heart: titled, a cavalryman (dragoons) and a keen sportsman. *Rittmeister* Erich *Graf* (Count) von Holck was also impatient and a trifle eccentric – he always took his small dog up with him; it lay on a fur rug on the floor of his cockpit. He had made a long rail journey to join his new unit. When the train stopped some fifty kilometres from his destination and the delay dragged on, Holk left his orderly to look after the luggage and the dog while he walked ahead, intending to board the train again when it overtook him. It never did: he arrived at the airfield twenty-four hours before it. Manfred's comment was that he was so fit that the long tramp had little effect on him. He also rated him highly as a pilot, praised his determination and said that not only did he have rare talent, but also he himself never had a greater feeling of security with so young a pilot.

Their last flight together nearly ended in disaster and was a valuable lesson for both of them in thermo-physics and aerodynamics. They had not been briefed to reconnoitre any particular area, and such liberty to use one's initiative was, Manfred thought, 'one of the nicest things about the air service'.

The Russians were retreating fast, leaving burning fields, woods, homesteads and villages behind them, with the Germans in pursuit. The most useful information Manfred could gather was the size of the beaten force and the axis of its march to the rear. Holck was flying him back to base at 1,500 metres altitude; ahead lay Wicznice, on fire, with dense smoke rising to an estimated 2,000 metres. Manfred, in the front cockpit, signed to the pilot to go round it, which would have added only five minutes to their flight time, but the impatient *Graf* had other ideas. Perhaps fortified by the knowledge that his blood was even bluer than his aircraft captain's (he was equivalent to a British earl), and certainly motivated by his delight in taking risks, Holck disobeyed. Reminiscing two years later, Manfred wrote, 'The greater the danger, the more it attracted him. It was fun to be with such a daring fellow.'

They flew straight into the choking, blinding smoke, ignorant of the effect that the super-heated air would have on the surrounding atmosphere and the engine. The Albatros stalled, spun and lost 1,000 metres before it emerged from the maelstrom and its pilot managed to regain straight and level flight. The engine began to falter, machine-guns on the ground opened fire on them, then the engine stopped. If they fell into Russian hands they would meet a savage death. The machine crashed near an artillery position from which the enemy had been driven the previous evening. Baron and count sprinted towards a nearby forest. Soldiers, whom they thought were the enemy, came running towards them from that direction; and turned out to be members of a Guards regiment commanded by royalty: the *Kaiser*'s second son, Prince Eitel, who provided them with horses on which to ride back to base, leaving their written-off aeroplane as a memorial to crass stupidity and indiscipline. The only casualty was Holck's dog – missing, believed deserted. He had lifted it out of the wreckage, unhurt but no doubt confused, and had to presume that it had run off with the soldiers.

Activity in this area was slackening and Holck was transferred to the Western Front. Manfred followed, to a different unit, on 21 August. On the way, his parents met him at Schweidnitz and took him and his orderly home for the night.

He described the enjoyment he derived from harassing the Russians, whom he despised.

> 'It was particularly amusing to pepper the gentlemen down below with a machine-gun. Half-savage tribes from Asia are much more startled when fired at from above than educated Englishmen. It is particularly interesting and amusing to fire at hostile cavalry: they rush in all directions'.

He does not seem to have felt any pity for the horses he professed to be so fond of but apparently killed and maimed without regret.

In the interests of security, his new posting was called Carrier Pigeon Unit Ostend. It was stationed at this seaside resort, but had nothing to do with homing birds: its work was bombing and reconnaissance. It flew two types of aeroplane. In the Albatros CI, the pilot sat in the front cockpit and the observer in the rear, where he had a ring-mounted Parabellum machine-gun that covered both sides and astern of the aircraft and could fire upwards. Their view was improved by a dual-curve cut-out in the upper wing trailing edge and rectangular cut-outs in the lower wing roots. This type of aircraft was used for photographic or visual reconnaissance, artillery observation and light bombing.

A bomb load of 70 kg could be stowed vertically in a space between the cockpits. The engine was a 160 hp Mercedes DIII. The aeroplane was 25 ft 9 in long, its maximum speed was 87 mph and its service ceiling 9,840 ft.

Manfred recorded that he had been 'fascinated by the name *Grosskampfflugzeug*', large battle aeroplane, which the unit also flew. The first type thus called was the AEG G II, which appeared in July 1915. Powered by two 150 hp Benz Bz III engines, it carried a machine-gun in the front cockpit and a 200 kg bomb load. Some 30ft long, it could attain 85 mph and 12,000 ft.

Georg Zeumer had joined the squadron some time previously and Manfred was delighted to be able to fly with him again. They spent five or six hours a day in the air. Their first encounter with the Royal Flying Corps, on a sortie in an Albatros C I, was with a Farman that they thought would be easy meat. Some variants of this primitive-looking two-seat pusher reconnaissance machine had been fitted with a Lewis gun for'ard. Its best speed was 65 mph. The Germans attacked and the two aircraft rushed at each other head-on. Manfred managed to fire four rounds, but missed. The Farman turned onto its opponent's tail and gave it a good burst but also did no damage. After they had chased each other in circles for a while, the Farman departed. The two Germans blamed each other, but eventually agreed that to be fully effective against a target in front the pilot needed a gun, so as to co-ordinate positioning the aeroplane with aiming, and the observer should defend against attack from astern. They did not appreciate that it was also a display of a fact that fighter pilots in both World Wars had to learn: an aircraft slower than its adversary could make this handicap an advantage by turning inside it.

Manfred's pleasure when bombing was marred because he could not see the bombs burst, as it meant looking astern and the view was obscured by the aircraft's wings. One day, after bombing, he gestured to Zeumer to turn and bank so that both could enjoy the spectacle of high explosive blowing up British troops. The propellers on either side were so close to the front cockpit that one sliced the tip off Manfred's little finger.

He had bought a pure-bred 'elm-coloured Great Dane for five marks from a nice Belgian'. He named it Moritz and took it for a flight – once. 'He behaved very sensibly and looked around with interest, but my mechanics were angry because they had to clean the aircraft of some unpleasant things.'

In the third week of September, three of the unit's Albatros C Is and crews were detached to operate over Champagne, where they found that two Fokker E Is were also based. The latter was the best German single-seater at the time. Zeumer was at once besotted by it and spent as much time as possible flying it. Capable of 80 mph and able to climb to 10,000 ft in forty minutes, its greatest assest was that it was fitted with an interrupter gear, which permitted its machine-gun to fire between the blades of the turning propeller.

This was the second single-seater to be able to fire through the propeller disc. The first had been the invention of a French pilot, Lieutenant Roland Garros. After much experimenting, he had achieved success by fitting wedge-shaped steel deflectors to the blades, so that all bullets that did not happen to pass between them would be turned aside instead of removing lumps of wood and wrecking the airscrew. However, this reduced the number of bullets that were effective. When Garros shot down his first enemy aircraft on 1 April 1915 and four more during the next two weeks, he revolutionised air combat. On 18 April, on his way to bomb the railway station at Courtrai, he flew so low that one rifle bullet from a German soldier smashed his petrol pipe. Not only was this an ignominious way to be brought down, but it also mean that his invention was revealed to the enemy. The propeller was sent to Anthony Fokker, the brilliant .Dutch aircraft designer. Early in the war he had offered his services to the British, but been rejected. He now worked in Germany: five weeks later he had perfected an interrupter gear for the E I. One unavoidable effect was that it slightly slowed the rate of fire; the other was that its timing was sometimes faulty and pilots, including the two most brilliant at that time, Boelcke and

Immelmann, sometimes shot their own propeller blades to bits.

Neglected by Zeumer, Manfred began flying with another highly experienced pilot, *Oberleutnant* Paul von Osterroth, a pre-war regular, who had joined the air service immediately on completing his cadet training. Shortly after the brief fight with a Farman, they spotted another of the same type about five kilometres inside the French lines. Its occupants were oblivious of danger while Osterroth approached close enough for Manfred to start shooting. He fired 100 rounds before his gun jammed, but that was enough to send the French aeroplane spinning down, to crash in a shell crater. This was never credited to him as a victory, because the High Command acknowledged only aircraft shot down behind German lines and confirmed by independent German witnesses.

As October began, there was more redeployment of the flying units. Manfred returned briefly to Ostend before he was transferred to Rethel, on the River Aisne, 135 miles south-west. The unit had its own train, so that the whole complement of air and ground crews, luggage, equipment, aircraft (wings detached from the fuselage) and spare parts could be moved quickly from any part of the front to another. This journey had a radical effect on his future.

Germany's two leading fighter pilots were *Hauptmann* (Captain) Oswald Boelcke, with four victories, and *Oberleutnant* Max Immelmann, with five. Since April 1915 they had been in the same unit, No. 62 based at Douai, which was equipped in May with the Fokker EI. They shared a house and had common characteristics: both were thorough and methodical, treated air fighting as a science, systematically worked at its basic principles and arrived at the best ways of applying them. As a result of their discussions, Boelcke propounded a doctrine that, in essentials, remained valid through all the decades during which single-seater fighters fought each other with machine-guns or cannon.

Boelcke's fighter pilot's gospel began with the basic essential that, when opening an attack, he should have a great height advantage and the sun behind him.

The rest of his principles were:

43

- He should make use of cloud to conceal his approach.
- His objective should be to get so close to his target that he could not miss it, yet be in a position where the enemy could not bring a gun to bear on him. In a fight between two forward-firing single-seaters, this means being astern, and slightly above, the other aircraft. A single-seater fighting a two-seater that has a rear machine-gun needs to position itself behind and slightly below.
- If attacked from ahead, a pilot should turn directly towards his adversary. This presents the smallest target and reduces the time in which the other man can aim and shoot.
- If attacked from behind, a pilot should bank into the tightest possible turn. This makes it difficult for the enemy to get on his tail; and, if he turns inside his enemy, he has a chance to get on *his* tail.

Boelcke, with his greater number of flying hours and more mature character than Immelmann, was the world's first great fighter tactician.

He belonged to a different unit from Manfred but was on his way to Rethel aboard the same train. In the dining car, Manfred recognised him from photographs he had seen in newspapers. It was then that he asked him to explain the secret of his success and received the half-jocular reply about getting very close to the target before shooting. For the rest of the journey Manfred courted Boelcke's friendship: they played cards and talked for many hours until eventually his new friend advised him to learn to fly a Fokker in order to improve his chances of shooting the enemy down.

In October, the Germans had equipped three units, among them the one that Boelcke commanded and on which Immelmann served, with the the Fokker E2. This, with its 100 hp Oberursel engine that gave it 100 mph and a 12,0000 ft ceiling, which it could reach in thirty minutes, had entered service in July. While Immelmann stayed at Douai, Boelcke went to Rethel to take part in the newly introduced barrier patrols: singletons prowling along the Allied front, waiting to pick off enemy aeroplanes with the E2's now established dive out of the sun. He was operating in the French sector.

As soon as Manfred's unit was installed at Rethel, he pestered Zeumer to teach him to fly in the spare time when he himself was not on sorties in the Albatros C I and Zeumer was not flying the Fokker E I. There was a dual control aircraft on the aerodrome, in which he began his instruction. By the 10th of October, his instructor deemed him ready for his first solo. Manfred was not so sure, but had to accept his friend's opinion. All went well until the landing, when he made a common error of judgment and, as the RAF would say in a later war, 'bent' the nose and undercarriage. His next flight was completely successful and he confidently underwent a proficiency test two weeks later, which entailed taking off, describing an accurate figure of eight, and landing, five times. He failed.

On 15 November he was sent to the flying school at Döberitz to train as a fighter pilot. For relaxation he used to fly to a friend's estate, where he could land, to shoot wild pig. Sometimes he stayed overnight for a moonlight hunt. He passed his final examination on Christmas Day and was able to get home that evening, where his sister and brothers had already assembled with their parents. For nearly three months more he had to continue flying practice when the weather allowed before he was considered proficient enough to join a squadron.

Chapter 6

The Fighter Pilot

On 16 March 1916 Manfred reported to *Kampfstaffel* (Battle Squadron) 8, whose establishment was six aircraft. It was stationed at Metz and, flying the Albatros CIII, which was slightly faster than the CII, operated on the Verdun front against the French. Command of two-seat fighters had passed from the observer to the pilot, but this was not enough for Manfred, who wanted to shoot at the enemy, too, not merely position the aircraft so that his observer could do so. He copied what some RFC pilots had first done in 1914 and many Frenchmen adopted later he had a machine-gun, pointing forward, mounted on the upper wing. It was not until 26 April that he was able to use it. He sighted a Nieuport 11 below, a nimble single-seater on which a Lewis gun similarly mounted was standard. He dived, took the pilot by surprise and shot it down. It landed in a wood behind the French lines, so, once again, he was denied credit for a victory; however, it was mentioned in the official report for that day, although his name was not stated. He wrote at once to give his mother 'the good news'.

Boelcke had returned to Douai in December. He and Immelmann entered the new year with seven victories each. On 12 January 1916 Boelcke won his eighth and Immelmann caught up with him the next day. They both held the Iron Cross, First Class and now received their country's highest decoration, *Pour le Mérite*, known as the Blue Max for its colour. It bore a French name because this was the only language that Frederick the Great, King of Prussia 1740–1786, who instituted it, could speak.

In the eyes of German fighter pilots, these awards set a precedent and they all strove to reach eight victories, in expectation of being similarly honoured.

An arbitrary standard of excellence among fighter pilots had first been set by French journalists. Before the war, they used the term 'ace' to describe Adolphe Pégoud, who was acknowledged to be France's best pilot. He was undoubtedly the most experimental and among the bravest: the first to fly inverted and the first to perform an outside loop, which became known as a bunt. This is so dangerous that it has always been forbidden in the RAF, except for test pilots when flying specific types. He was shot down and killed in 1916.

Air fighting, because of its novelty and the false glamour that reporters attributed to it, made better reading than ground battles. Likening fighter pilots to armoured knights who fought each other to the death with lance and sword, but chivalrously spared a wounded and unhorsed opponent who had fought well, was a romantic absurdity. French journalists who extolled their own fighter pilots' achievements elevated all who scored five kills to the status of ace. The *Aviation Militaire* did not at first recognise this, but tacitly had to accept it. In time the term insinuated itself into official use. *Commandant* Tricornot de Rose, who commanded the French squadrons on the Verdun sector, seeing the value of it in maintaining morale, duly began to publish his pilots' scores.

The RFC disapproved of personal publicity and discouraged it. The British, with their ideal of team work or play and modesty about individual feats, approved only of praise directed at a squadron. Most commanding officers did not like having brilliant individualists in their units. As the war advanced, fighters fought in pairs, threes, four, sixes and whole squadron strength of twelve. It was also regarded as inviduous to elevate the value of fighter pilots' work above that of bomber, reconnaissance and artillery observation air crews.

The adulation of outstanding British fighter pilots began in 1916, when the public first read in newspapers about nineteen-year-old Lieutenant Albert Ball, a religious youth who was as ruthless as Manfred von Richthofen. He had joined No. 13 Squadron in February to fly BE2cs, which were outclassed by almost every other type at the Front. Although the squadron's main task was artillery spotting, in April he enabled his observer to shoot down an enemy aircraft and forced two others to land.

In May he was posted to 11 Squadron, which had eight FE2bs, four FB5s and three Bristol Scouts: all inferior to the enemy fighters that preyed on them. However, the squadron was being re-equipped with the Nieuport 11, which performed better than the Fee, Gun Bus and Bristol Scout. It was armed only with a Lewis gun on the upper mainplane, but this had the new Foster mounting: a curved rail along which the gun could be slid down towards the cockpit to make reloading easier. This was still a primitive arrangement compared with the interrupter gear that the Germans and French were using. The British were trying out two similar mechanisms. The first was the Vickers-Challenger, which had been tested with a Vickers gun on Royal Naval Air Service Sopwith 1½-Strutters, two-seat reconnaissance fighters – an apparent contradiction in terms. This was soon replaced by the better Scarff-Dibovski gear.

Ball's letters to his parents reveal him as immature and naïf, but he displayed the cunning of a born stalker and superb marksmanship, although his upbringing was diametrically different from Manfred's.

On 10 July he wrote home,

'You ask me to let the devils have it when I fight. Yes, I always do let them have all I can, but really I don't think them devils. I only scrap because it is my duty, but I do not think anything bad about the Hun. He is just a good chap with very little guts.'

This accusation was a wild generalisation, for the average German airman did not lack courage. He continued, 'Nothing makes me feel more rotten than to see them go down, but you can see it is either them or me [Trent College had evidently neglected his grammar] so I must do my best to make it them.'

Flying a Nieuport 11 he proved his exceptional skill by

shooting down an Albatros AI. In July he joined 60 Squadron. His twentieth birthday fell on 14 August. By the end of that month he had flown eighty-four sorties shot down sixteen aircraft and destroyed an observation balloon.

The Foster mounting was convenient but not perfect. On 15 September his combat report stated that after he had fired one drum from fifty yards at an Albatros reconnaissance type, 'the gun on the Nieuport came down and hit me on the head, preventing me from following the HA* down.'

He never hesitated to take on several opponents at a time, single-handed. On 21 September he met six Rolands flying at about 90 mph. His customarily impersonal combat report, which reads as though the machine were flying itself, says:

'HA seen N of Bapaume in formation. Nieuport dived and fired rockets.† Formation was lost. Nieuport got underneath nearest machine and fired a drum. HA dived and landed near railway. Nieuport then attacked another machine and fired two drums from underneath. HA went down and was seen to crash at side of railway. After this the rest of the HA followed the Nieuport towards the lines and the Nieuport turned and fired remainder of ammunition after which it returned to the aerodrome for more. Second machine was seen to crash by Lieut Waters.'

On 25 September:

'Two formations came along, Nieuport attacked the first. The HA ran with noses down, but, when another formation came near it turned towards the Nieuport. The Nieuport fired one drum to scatter the formation after which it turned to change drums. One of the drums dropped into the rudder control and for a few seconds the Nieuport was out of control.

'Nieuport succeeded in getting drum on gun and attacked

---

* Hostile Aircraft
† Le Prieur, electrically operated, five attached to the wing struts on either side.

an Albatros (two-seater) which was then flying at its side. Nieuport fired 90 rounds at about 15 yards range underneath HA. HA went down quite out of control and crashed. The remainder HA followed Nieuport, but in the end left. In order to keep them off at a safe range, Nieuport kept turning towards them. Each time this was done HA made off with noses down.'

He might have fired those ninety rounds without changing drums: in the spring of that year Lanoe Hawker had introduced a double drum – two forty-seven-round ones welded together.

By now Ball had been promoted to captain and awarded the Distinguished Service Order and Military Cross.

His fighting style makes an interesting contrast with Manfred's.

It was not only in action that Manfred displayed occasional rashness. During the spring and early summer of 1916 there were several thunderstorms over the Verdun front. One day he had to fly from Mont, where he was stationed, to Metz on business that did not involve an observer, so went alone. As he prepared to return to base, a thunderstorm was brewing. He wrote later, 'I had never tried to fly through a thunderstorm, but could not resist the experiment'. Several experienced pilots advised him not to take off, but he ignored them because he thought he would seem timid if he heeded their warnings. His route took him across the Moselle mountains and through valleys where a high wind was causing tremendous turbulence, rain was hammering down and lightning was sizzling around him. 'It was like riding a steeplechase over trees, villages, church towers and rooftops. I believed that death could come at any moment', but added that there had been beautiful moments to compensate for the fright.

His greatest desire was to fly single-seaters. He constantly pressed his squadron commander to let him do so, until eventually he agreed to allow him and another pilot, Hans Reiman, to share a Fokker: Manfred to make the morning patrols and Reiman the afternoon ones. On their second day, Reiman, fighting a Nieuport, was forced down in no-man's-land.

He set fire to his machine and, after dark, made his way to the German trenches.

The Fokker, despite its many good qualities, had a notoriously unreliable motor, which made its pilots reluctant to cross the British or French lines. When a replacement aeroplane was delivered, Manfred's first sortie was drastically cut short: the engine failed just after take-off, he crash-landed and the brand new fighter was badly smashed. At every flying school in the world, one of the first orders pupil pilots were given was never to turn back if they lost an engine: the aircraft would inevitably stall into a fatal crash; they must always land straight ahead, however rough or tree-encumbered the terrain. Many, neophytes and the experienced, who ignored this had proved the literal truth of the warning. Manfred was wise not to succumb to his habitual over-confidence and relish for taking risks and try to regain the aerodrome.

On 18 June, Immelmann was killed in combat with an FE2b of 25 Squadron, to the pilot and observer of which the RFC credited the victory. The Germans, on the contrary, insisted that Immelmann had shot off his own propeller again. They also misidentified the British aircraft as a 'Vickers two-seater', thinking it was an FB5 Gunbus; but then, they also referred to the de Havilland DH2 as a Vickers.

Although Immelmann had been assessed the best pupil on his flying training course, he broke several aeroplanes during his early career. His most important contribution to fighter tactics was the turn he invented, which is still known by his name. Its purpose, if his first attack did not despatch the enemy, was to regain height and make a second attack in the shortest possible time. After diving and firing, he continued the dive below his target, then began a loop that would take him climbing fast ahead of it. At the top of the loop, he rolled from an inverted to an upright position and ruddered hard into a diving turn that brought him down on the other aircraft's tail once more, and fired again.

His death was a demoralising shock not only for the flying service but also the whole nation. The Kaiser, determined not to risk losing Boelcke as well, sent him on a public relations tour of Germany's allies in the Near East and Balkan countries. This was a disappointment for Manfred, who was on the point of asking for a posting to Boelcke's squadron.

In the same month, a new offensive began on the Eastern Front. At the turn of the year the German Military Air Service had been reorganised and formations, known as *Kampfgeschwader*, the equivalent of British Wings, were formed. The one to which Manfred's squadron belonged, No 2, was transferred to the Russian Front in July. There, the personnel lived aboard the train that had brought them, but the nights were so airless and hot that he and two other pilots camped in a forest beside the track, in which they also went hunting during the night.

The Albatros CI returned to its bombing role, which he found himself enjoying. The Russians had few aeroplanes, so the chances of shooting one down were slim, but by dropping bombs he was striking hard at the enemy and relished the spectacle of the damage he was doing. Having bombed, he used to go down to strafe whatever ground troops he could find and said that it was most entertaining to attack a cavalry column, because it threw horses and riders into such confusion. Did he ever, as a lover of horses – or was it only of riding them? – feel remorse and pity for those he killed and wounded?

General ignorance about flying prompted many ridiculous rumours on all the battle fronts. One, derided in an article by C.G. Grey, Editor of *The Aeroplane*, was about a German aircraft which climbed so high in evading its Russian attackers that both the occupants froze solid. The aircraft was alleged to have glided down to a landing behind the Russian lines, undamaged. The pilot and observer were found 'untouched by bullets, but stone dead with the cold'!

Changing fortunes at the Western Front necessitated another revision of German fighter units and the introduction of the name *Jasta*, an abbreviation of *Jagdstaffel* (fighter squadron). In mid-August Boelcke was recalled from Turkey, which he had reached on his foreign tour, and given command of *Jasta* 2. He resolved that it would be an élite body and set about selecting the pilots. Two whom he chose were Manfred and a thirty-seven-year-old *Leutnant*, Erwin Böhme.

Chapter 7

The Real Start
of a Famous Career

On 1 September 1916, Manfred arrived at his new squadron's airfield at Bertincourt, near Cambrai. Since the first few months of war, when no British or German aeroplanes mounted a machine-gun and pilots and observers began to carry rifles or shoot at each other with revolvers, the war in the air had passed through two major phases and was entering a third.

During the initial period of bumbling attempts by the Royal Flying Corps, *Aviation Militaire* and *Luftstreikräfte* to shoot one another down, the British often and the French sometimes fitted a Lewis machine-gun to the upper wing of their biplanes, a rough and ready improvisation. By the time war was declared, France was manufacturing more aeroplanes, of more types, than any other European country. It was not surprising, therefore, that when the war began one of those in military service, the Voisin III (Type L), a pusher two-seater in which pilot and observer sat side by side, had a Hotchkiss machine-gun in front of the latter on a properly designed mounting. The advantage this could give over the enemy was made obvious when, on 5 October 1914, one of these, flown by Lieutenant Joseph Frantz, with a mechanic, Private Louis Quénault, manning the Hotchkiss, shot down a German Aviatik two-seater, the first flying machine ever to be brought down in this way. It is remarkable that no use was made of these aircraft, with their evident potential for wreaking mayhem on a large scale, as a fighter. One reason appears to be that owing to the plethora of aeroplane types and the many design weaknesses in those early years, manufacturers were kept so busy making spare parts that production of complete machines lagged.

The more important reason was that the Voisins were regarded as essentially bombers. In 1914 the French had already perceived the importance of not only tactical but also strategic bombing. The First Bomber Group, comprising three Voisin *escadrilles* was formed. When the Germans fired chlorine gas shells into the French lines for the first time, on 22 April 1915, they invited a terrible reprisal. On 26 May all three Voisin squadrons bombed the acid and chlorine works at Ludwigshaven and Oppau, an operation that involved five hours' flying. The priority given to bombing meant that no Caudrons could be spared for fighter duties.

Quénault's triumph was not the first aerial victory. On 13 August 1914, Lieutenant H.D. Harvey-Kelly of No. 2 Squadron, which flew BE2cs, was the first to take off for France. It was fair enough, therefore, that on 25 August he and two others were the first in the world to bring down an enemy aircraft, although they did not fire a shot. How could they? Their aircraft were unarmed. Their escapade was a typically mischievous, high-spirited RFC (and RAF) prank. Harvey-Kelly had intercepted one of the earliest German machines, a wretched affair called a *Taube* (Dove), and proceeded to torment its two occupants by flying a few feet astern and following its every evasive movement. Presently two of his squadron mates joined in the fun, one on each side of the quarry, and the three of them forced it to land. Its pilot and observer bolted into a wood where the three Britons pursued but did not catch them, so set fire to the abandoned aircraft.

The practice of lashing a Lewis gun somewhere and somehow within the pilot's reach, instead of fixing it firmly in the cockpit and under his hand, did not end with Roland Garros's innovation of bullet-deflectors on propeller blades. When the Germans swiftly improved on the device in the winter of 1915, the Fokker E series, the first real single-seat fighters, began to enjoy a period of supremacy on both the British and French fronts. The EI was soon followed by the faster EII. Both were highly manoeuvrable and difficult to see head-on, being monoplanes with thin wings and a slim, round fuselage. The Vickers FB5 Gunbus could not out-manoeuvre them. Although the new Nieuport 11, known as the *Bébé*, was as agile as the Fokker, it carried a Lewis gun on the upper wing, which, even with a purpose-designed mounting, was much inferior to a Spandau in front of the cockpit. No British or French fighter could match the German's fire power. The RFC's casualties were so heavy that Parliament debated the 'Fokker scourge' and air crews were spoken of as 'Fokker fodder'.

The second phase began in February 1916 with the formation of the first homogeneously-equipped British single-seater fighter squadron, No. 24, and the entry into service of its aircraft, the DH2. This was a pusher, its only weapon a Lewis gun mounted in front of the pilot, but it was more manoeuvrable than the Fokker E I, II and III. No. 24 Squadron began to develop offensive rather than defensive tactics when escorting reconnaissance and bomber machines as well as patrolling in search of enemy fighters. Some RFC squadrons were being supplied with the Nieuport 11, which was also winning fights against the Fokker, thanks to its agility. A new British fighter that matched the Fokker in performance but was handicapped by having a wing-mounted Lewis gun also entered the lists in August: the single-seater Sopwith Pup, with a top speed of 105 mph.

The third phase was introduced by the advent of the Albatros DI and II, their speed and armament both superior to any other fighter's in the world.

During the Verdun battles the French air force had introduced another innovation, this time in fighter organisation. Its first élite *escadrille* was N3 (for Nieuport). It was known as *Les Cigognes*, (The Storks), because each aeroplane bore on both sides of the fuselage a depiction of this bird in flight. The unit was expanded into a *Groupe* of six Nieuport *escadrilles*, N3, N23, N26, N73, N103 and N167. *Capitaine* Felix Brocard, N3's commanding officer, now became the first commander of a fighter Wing and led the first large fighter formations.

The popular image of a fighter pilot of any nationality had by now become firmly established. A French magazine, *La Guerre Aérienne*, described him as 'a man exceptional for his physical and moral qualities, an adventurer out of the ordinary; a sort of champion towards whom popular fervour is directed'. This was why, it said, everyone who entered the aviation service aspired to fly a fighter.

The first Frenchmen to attract fulsome praise were Jean Navarre and Charles Nungesser, on the Verdun Front. By April 1916, Navarre had shot down seven enemy aircraft and Nungesser had scored six.

According to *La Guerre Aérienne*,

'Fighter pilots are an élite, a glorious élite, universally praised, officially very much appreciated. To become one of them is to receive a mark of distinction, it is the consecration of exceptional qualities. They are entrusted with a supple, highly-strung prodigiously fast machine. This seduces sportsmen by its lively charms, in particular the intoxicating speed that increases tenfold the sensation of power. What justifies the fighter pilots' liberty is the fact that they put it to good use. For that intrepidity, courage, love of sport and a taste for risk are necessary.'

The magazine even praised the 'daring, even temerity' of the British.

'Every day they accomplish exploits that prove how useful sport is as a training for those who make war. Perhaps one could even reproach them for their sporting spirit, which makes them ignore danger. They hurl themselves at the enemy impetuously, with admirable bravery.'

Reference to sport is erroneous in the context of a lethal contest. The French, who assembled their best pilots in virtually segregated élite units, did form a squadron, N77, which enrolled noted sportsmen. One had captained France at rugby football, others were international horsemen, fencers and racing drivers. They were not conspicuously more successful than others. It is undeniable that young men who pitch themselves whole-heartedly into sports and games, particularly those in which injuries through bodily contact are common, show the kind of spirit that makes good fighters on land or at sea as well as in the air; but that is as far as the association of sport with war goes.

Manfred von Richthofen possessed this spirit abundantly. He fought with the bravery and dash that he displayed in equestrian competitions, which were as dangerous as any games that the British or French played.

Certainly, the RFC liked to play games for recreation when possible. This is evidenced in *War Flying*, written under the pseudonym 'Theta', and published in November 1916, containing letters home from an 18 year-old pilot who joined a BE2c squadron in France that year. He wrote,

'Everyone here is cheerful and thinks flying is a gentleman's game and infinitely better than the trenches; when your work is over for the day there is no more anxiety until your next turn comes round, for you can read and sleep out of range of the enemy's guns . . . There has been a craze here for gardening recently and people are sowing seeds sent over from England and building rockeries and what not . . . There is a river near the aerodrome where we all go swimming . . . The ping-pong (table tennis) set has arrived . . . Cricket is the great "stunt" here in the afternoon and rugby in the evenings. The mornings are spent repairing the damage of overnight caused by rugger. All this, of course, provided the little incidentals of flying and so on do not interfere to excess . . . We have a game here now which is something like tennis. Instead of racquets and balls we use a rope quoit which must be caught and returned as per tennis. We are getting a tennis court made . . . Yesterday I was in the middle of a game of tennis when, with one or two others, I was ordered to fly over to a neighbouring aerodrome to be ready for a special job in the morning.'

Manfred and his squadron comrades had no time for such pursuits; anyway, sport meant, to them, shooting and riding. They were single-mindedly dogged about killing the enemy. There is no record of the French sportsmen's squadron indulging in games at the front, either.

Adulation of fighter pilots, in less ecstatic terms than the French journalists', was accorded by the German press to

Boelcke, Immelmann and other celebrities such as Gontermann and von Röth, who specialised in destroying observation balloons, which were strongly protected by anti-aircraft batteries. They both won the Blue Max. The awards of British decorations were, of course, published in the home newspapers as well as the official *London Gazette*. For instance, Major Lanoe Hawker, commanding 24 Squadron, was well-known because he had won the Victoria Cross and Distinguished Service Order in 1915. With so much publicity, both sides knew the names of their most dangerous opponents.

News came also of events on the Italian front, where Italy was fighting Austria-Hungary. Francesco Baracca, destined to become his country's leading ace, Fulco Ruffo di Calabria and Pier Ruggero Piccio were becoming known to a wider public than their native one.

All this was heady stuff for the contestants in all the air forces. Manfred, who had been raring to be embroiled where the hottest air fighting was to be found, was keyed up to show what he could do when provided with a first-class aeroplane, a devastating pair of guns and an abundance of opponents. So were his squadron comrades.

The challenge to Boelcke's *Jasta* was uncompromising and all the more provocative because of the manner in which the RFC had won the upper hand over the hitherto invincible Fokker E series. On 24 April, four of 24 Squadron's DH2s were escorting five BE2cs of 15 Squadron on reconnaissance. Twelve EIIIs took them on, some circling to prevent the BE2cs turning back, the rest waiting to make their familiar dive and zoom. When it became obvious that the little reconnaissance machines, as defenceless as sheep against wolves, had no intention of going home yet, by which time the whole British formation was deep behind enemy lines, the circling Fokkers re-formated on the others and the whole lot swarmed down on the British. The tightness with which the DH2s turned and made straight for their attackers came as a surprise. Disconcerted by this unprecedented tactic, the Germans pulled out of their dives and broke right and left. Within seconds the DH2s were among them and shooting. Three Fokkers quit the scene, damaged. The remainder withdrew to circle again, hoping to draw the defenders away from their flock, but Lanoe Hawker's pilots stayed put. The Fokkers did not attack again and all nine British aeroplanes went home unharmed.

The DH2 had broken the Fokkers' grip on the Western Front and, five months on, it was the intention of *Jasta II*'s commanding officer and all his pilots to re-establish German supremacy at once.

The manner of Manfred's first victory for the *Jasta* has already been told. The significant psychological change he experienced during that combat radically affected his attitude and skill. He had begun the fight wondering whether he would win. Presently he told himself that his opponent '*must* fall' and this absolute determination, to which he attributed great importance as essential total resolution in any fighter pilot, was to be his mental attitude in every future contest. In this engagement Boelcke shot down his twenty-seventh victim.

Six days later Manfred shot down a Martinsyde, which was like destroying a 1924 Hawker Woodcock (top speed 132 mph, twin Vickers guns) with a 1940 Messerschmitt 109E (354 mph, various combinations of machine-guns and cannon). The helpless pilot, of 27 Squadron, was killed. Manfred again ordered a cup to celebrate the event and continued to do so after every kill. He took the Martinsyde's gun as a souvenir.

A week after that he brought down an 11 Squadron FE2b and killed the pilot and observer. This was a slightly more creditable performance, but the odds were still heavily in favour of the Albatros. For a souvenir this time he cut away a piece of fabric that bore a roundel.

Henceforth, unless the aeroplane he had sent down was burned to a cinder or souvenir hunters got there first and stripped it, he always took or had someone else remove some portion of it to send home, where his family vulgarised a room dedicated to exhibiting such relics and his self-awarded silver cups. The machine-gun was, in deplorable taste and self-advertisement, displayed over the front door.

His next three successive victims were BE2s, undistinguished single seaters, two of 21 Squadron and one of 19 Squadron, on 7, 16 and 25 October. All the pilots were killed.

On 28 October Boelcke's career came to an end in the most unexpected way and with potentially grievous effects on the *Jasta*'s morale. Two 24 Squadron pilots, Lieutenants Knight and McKay, were far behind enemy lines when Boelcke and his *Jasta* intercepted them. The RFC's combat report says that twelve Halberstadts and two small Aviatik scouts attacked a pair of DH2s. It was the British pilots who misidentified their assailants this time: the Albatros D rather resembled the Halberstadt, although to a well trained pilot it had obvious differences, such as the shape of the tail fin and less shark-like lines of the fuselage. The two Aviatiks were presumably on a bombing raid with fighter escort. The DH2s began to circle tightly. Boelcke and Böhme together attacked one of them but Manfred, chasing the other, crossed their path. Böhme's left wing hit Boelcke's right and Boelcke's machine went down steeply, out of control. The damaged wing came off and he was killed on impact with the ground. He had 31 victories to his name.

The RFC dropped a wreath at his funeral, with a note reading, 'To the memory of Captain Boelcke, our brave and chivalrous foe, from the British Royal Flying Corps'. This was a tribute to his habit of visiting hospitals where men he had wounded were being treated and of giving or sending them cigarettes.

In a letter to his mother, describing the crash and his part in the funeral, at which he carried Boelcke's decorations displayed on a cushion, Manfred wrote,

'We are deeply affected, as if we had lost a favourite brother. The funeral was worthy of a prince. In the last six weeks we have lost six of our twelve pilots, killed, and one wounded. Two have suffered complete nervous breakdowns.'

This is hardly the information a considerate son would have given either of his parents.

Oberleutnant Stephen Kirmaier, the second-in-command, took over the *Jasta*. Five days after Boelcke was lost, he shot down his eighth enemy aircraft and, eight days later, his ninth.

On 3 November Manfred raised his score to seven with a victory over an FE2b, killing its occupants. On the 9th he destroyed a BE2c and mortally wounded its pilot. Both he and Kirmaier expected the Blue Max, now that each had bagged eight hostiles, but were disappointed.

On 20 November Manfred scored a double by sending down another BE2c, whose crew were taken prisoner; and an FE2b whose pilot died of wounds and observer was killed.

On the same day, Kirmaier had his eleventh victory. On the 22nd, fighting an alleged Vickers, he uncharacteristically followed it over the British lines and was killed. Command fell temporarily to the Senior Administrative Officer, but Manfred, in view of his personality and the fact that he was now the squadron's top scoring survivor, led it in action. He took a serious view of his new responsibility and began by explaining to his pilots the ways in which air fighting was changing and the tactics they must adopt in order to cope with them. Flying discipline had always been essential and one facet of this was good formation-keeping, in which he exercised them strictly.

On 23 November he had the most important victory of his whole career. The man he fought, Major Lanoe Hawker VC DSO, was not only a fine pilot but also as good a marksman as Manfred and, like him, a regular officer and scion of the landed gentry.

In the early days, he used to take a rifle up with him, resting against his right leg, even when in a BE2c, whose observer had a Lewis gun. As a flight commander in No. 6 Squadron, on 25 July 1915, flying a Bristol Scout, he destroyed two hostiles on the same sortie. Rather than have a Lewis gun on the upper wing, he was so confident of his accuracy that he had one mounted pointing to the left at an angle of forty-five degrees to his line of flight, so that its bullets would miss the propeller. Apocryphal stories about his amazing skill with a rifle had always abounded. The rumour that spread this time alleged that he had shot down three German machines with a rifle.

In September he had taken command of 24 Squadron, which was being formed in England and arrived in France on 7 February 1916. Hawker and his three flight commanders were the only ones with operational experience. He and one of these were aged twenty-five, most of his other pilots were twenty or younger.

When other squadrons came to dinner there were tumultous games in the mess, which became a tradition in RFC and RAF officers' messes and would never even have been contemplated in any other nation's. Battles were fought using armchairs as armoured cars, tennis balls as bombs, soda syphons as flame-throwers, and rugger was played to the detriment of furniture and windows.

Having been educated, after preparatory school, at the Royal Naval College, Dartmouth and the Royal Military Academy, Woolwich, he was a great believer in *Mens sana in corpore sano*: he took his pilots for a run every morning; borrowed cavalry horses for them to ride and had a tennis court laid out. The Germans, French and Italians, all of whom regarded the provision of brothels more important than facilities for exercise, would have regarded these means of providing recreation as wildly aberrant or puerile. The morale of his officers and men was high, they were delighted with him and respectful of his bravery.

Within a week of arriving at the front, two pilots had been killed when their DH2s spun into the ground. These tragedies were potential morale destroyers. Hawker took a DH2 up to 8,000 ft and, out of sight, spun it several times to left and right, with and without engine. When he landed, he told the pilots how to recover from a spin and they hurried into the air to practise it. Such a man was Lanoe Hawker, a light-hearted and far more attractive, many-faceted character than the serious, introspective Manfred von Richthofen.

When they met in combat Manfred won because, although Hawker was the more polished flyer, Manfred flew the faster, more heavily armed aircraft and his two guns were each loaded with 500 rounds, whereas Hawker's one had only ninety-four. Moreover, Manfred had had the great benefit of mock combat with a DH2 that Boelcke had forced down, intact, behind the German lines. Also, Hawker's engine was suffering from impeded petrol flow, which robbed it of full power.

Hawker, patrolling at 6,000 ft with Captain Andrews and

Lieutenant Saundby, had seen two enemy two-seaters, at which Andrews dived. The British combat report tells us:

'. . . and then, seeing two strong hostile patrols approaching high up, [Saundby] was about to retire when Major Hawker dived past him and continued the pursuit.

'The DHs were at once attacked by the HA, one of which dived on to Major Hawker's tail. Captain Andrews drove this machine off, firing 25 rounds at close quarters, but was himself attacked from the rear and his engine shot through almost immediately, so that he was obliged to try to regain the lines. He last saw Major Hawker engaging one HA at about 3,000 ft. Lieutenant Saundby could then see no other DHs and the HA appeared to have moved east, where they remained for the rest of the patrol.'

Manfred's combat report yet again misidentifies the DH2.

'I attacked in company with two aeroplanes of the squadron a single-seater Vickers biplane at about 3,000 metres. After a very long circling fight (35 minutes) I had forced down my opponent to 500 metres near Bapaume. He then tried to reach the Front, I followed him to 100 metres over Ligny, he fell from this height after 900 shots.'

The disparity in heights given by the reports was not unusual. Hawker, who had already been shot down twice by superior aircraft and wounded once, was this time fatally shot through the head.

Manfred helped himself to the DH2's machine-gun and a piece of fabric bearing the aircraft's serial number.

The fight had lasted more than half an hour because Hawker was a brilliant aerobatic pilot. Manfred, who despised this facility, never practised it and discouraged it in those he commanded, had never before met a master of the art.

Manfred was still leading the *Jasta* in action, but not experienced enough to warrant being appointed Commanding Officer. He and the other pilots applied for *Oberleutnant* Franz Walz to be their next commander. He was thirty-one and although he had only six kills to his name, they thought, from his reputation, that he had some of Boelcke's qualities. He arrived to take over on 29 November.

Meanwhile Manfred had notched up successes against two DH2s, with one pilot killed and one captured and two FE2bs, one pilot killed, one wounded. Thus 1916 ended with his total at fifteen. All had been achieved against aircraft that were at a disadvantage in speed, manoeuvrability or firepower.

On 11 December, the squadron was given the official title of *Jagdstaffel* Boelcke.

Chapter 8

Manfred's Attitude to Lothar

Over the years since the end of the Great War, Manfred has often been denigrated for jealousy of, and rivalry with, his brother Lothar, who was two years and four months his junior. On the contrary, the interest he showed in his brother's career can also be seen as affectionate concern and when, early in the war, Lothar was seeing more action than he, a wistful envy that was entirely proper. Manfred was a professional who had been trained for a military career from the age of eleven. War provided the opportunity for him to serve his country, do what he had spent eleven hard years preparing for and distinguish himself; yet for months he was kept on the fringes of the fighting. Lothar, who had been given a normal education before doing his compulsory military service, was in the thick of it. Disappointment and, for a Prussian cast in the military mould, humiliation, must surely have been the elder brother's emotions rather than resentment. Significantly, Manfred's detractors have all been of nationalities other than his own.

Both brothers were handsome, but physically and in character differed in many ways. Manfred was fair, short, introspective and reserved. However, in photographs with other members of his squadron he is usually smiling and has an air of geniality. Lothar stood a good half a head taller, was dark, broader-shouldered, sociable and cheerful.

Lothar *Freiherr* von Richthofen had joined the 4th von Bredow Dragoons, who took part in the invasion of Belgium. The regiment was soon transferred to the Russian Front, where the Germans won two great battles. The news in Lothar's letters

65

to his parents was passed on to Manfred, who was suffering agonies of frustration in the Verdun sector and wishing he had the same opportunities to do his duty in the fullest sense; and win medals.

From his earliest letters to their mother, Manfred had frequently stated that he was trying to win the Iron Cross. When he did receive it for his reconnaissances he was disappointed because it was only Second Class, not the First he coveted. When he was festering in a deep dugout, he wrote to her: 'Unfortunately, my regiment has been attached to the infantry, I say "unfortunately" because I feel certain that Lothar has already ridden in many cavalry charges, which we shall never do here.' This has repeatedly been quoted as proof of his jealousy, but it can just as logically be read as natural envy, not of his brother, but of the chance to do, himself, what he was brought up to do and to prove his manhood.

Transferring to the air service relieved his frustration and in February 1915 he persuaded Lothar, who was doing dull work training recruits, to follow his example. That does not seem like the advice of a brother jealous of his sibling, but the very contrary. As long as the younger man stewed in a non-combatant post, the elder, flying on operations, was in a position to feel that he had gone one better; if he were capable of such mean thoughts. The evidence is that such competitive feelings as he did have towards Lothar were frank and healthy. He encouraged him to fly and always sought his company. If he had wanted to indulge in 'one-upmanship' he would have been complacent about his young brother serving in an arm for which modern warfare had little use, not urged him to take to the air, where the prospects of hard fighting and fame were abundant. Lothar did not join the air service until some months later, but by the year's end was operating over the Western Front as an observer. In 1916 he qualified as a pilot.

Family ties between the six members of Major Albrecht *Freiherr* von Richthofen's household were close. Their affection for one another was manifested in many ways and communicates itself to anyone who studies their family relationships. On the second Christmas of the war, the Major and Lothar were home on leave, young Bolko was on holiday from his military academy and Manfred was able to join them late on Christmas Eve. Their unity and mutual love was clearly evinced. Manfred himself described the warmth and joy of the scene: his mother at the piano and the rest of the family grouped around her, singing, all in uniform including Ilse in her nurse's outfit.

The German sentimentality about Christmas is well known and was famously displayed on the war's first Christmas Day when the German soldiers took the initiative in shouting friendly messages from trench to trench and British troops mingled with them in no-man's-land.

In 1916 the Richthofens could not be united at home for the festival, but the father and two elder brothers were able to meet in *Jasta* Boelcke's officers' mess.

On the following day, Lothar, having returned to his flying training school, made his first solo. Manfred wrote to their mother to tell her of both events.

'Dad and Lothar were with me on Christmas Day. It was a memorable holiday and more fun than you at home might think. We had a Christmas tree and an excellent dinner. Next day, Lothar made his first solo flight.'

Manfred's pleasure in his brother's reaching this significant stage in his aspiration to qualify as a pilot is plain and demolishes all accusations of mean-spirited rivalry. There was, of course, competitiveness as both their flying careers advanced. This was true of all the pilots in the squadron and in the fighter squadrons of every nation fighting the war, and the Richthofens were open about it.

CHAPTER 9

THE PACE QUICKENS, THE DANGERS GROW

1916 had seen air superiority fluctuate between the Allies and the Germans, as well as many considerable advances in the performance of aeroplanes and armament. The airmen on both sides entered 1917 with expectation of further improvements in aircraft, engine and weapon design. All current machine-guns were prone to jamming. German pilots took a hammer up with them to free the mechanism of their Parabellums and Spandaus. Lewis rounds frequently jammed in the breech, which meant unscrewing the ammunition drum, a clumsy business; when the Vickers jammed, the belt had to be adjusted. All these interruptions meant either abandoning a fight temporarily or permanently, or being wounded or killed. The Germans often used incendiary bullets, so a high proportion of the aeroplanes destroyed caught fire and men were either burned to death or jumped out at several hundreds or thousands of feet to avoid this agony.

The recent emergence of publicised British aces, who were becoming as well known as the French and German, was an embarrassment to the War Cabinet as well as to the objects of hero-worship. Censorship could suppress neither falsehoods nor truths that were spread by word of mouth. The dominance of the Fokker EI and II over the British and French air forces had not been entirely concealed, for the newspapers published daily casualty lists. When the Royal Flying Corps and *l'Aviation Militaire* overcame the Fokker with the DH2 and the Nieuport 17 respectively, the British public was eager to know more about its leading fighter pilots than the names they read in newspapers when decorations were awarded.

Manfred gained his last victory of 1916 on 27 December, against an FE2b. The pilot was wounded, the observer captured. His first success of the new year was against a Sopwith Pup of the RNAS on 4 January. His report goes:

'A new type of machine, never seen before, but, as the wings were broken, barely identifiable. The Pilot, Lieutenant Todd [everyone was supposed to wear an identity disc], killed. Papers and valuables enclosed. About 4.15 pm. Just after starting we saw above us at 4,000 metres four machines unmolested by our artillery. As the anti-aircraft was not firing we supposed they were ours. When they were nearer we saw that they were English. One of them attacked us and we saw at once that it was superior to our machines. We detected its weak point only because we were three against one. I managed to get behind him and shoot him down. The machine disintegrated while falling.'

Having doubled the score for which Boelcke and Immelmann had been awarded the *Pour le Mérite*, Manfred was obsessed with anxiety about when he would be equally rewarded. His gloom at apparently being ignored, and being deprived by bad weather of the chance to add to his impressive record, was lightened when, on 7 January, some Albatros DIIIs were delivered to the squadron.

Exactly a week later, he was appointed to command *Jasta* 11, based at an aerodrome near Douai. He had to wait only two more days for the announcement that he had been given the Blue Max, which arrived during his farewell party. The newspapers fawned on him. Congratulatory letters, 'fan mail', came abundantly and he was pestered for portrait photographs.

He took his Albatros DIII with him and lost no time in setting an example to his twelve pilots. On 23 January he destroyed an FE8, killing its pilot, and on the following day brought down a FE2b, wounding both occupants, who were taken prisoner. About the latter encounter he wrote to his mother: 'One of my wings cracked at an altitude of 300 metres and it was a miracle that I reached the ground safely'. On the same day *Jasta* Boelcke lost three aeroplanes and it was found that two of them had suffered from the same structural defect.

The prevailing wind, blowing from the west, accelerated the British aircraft that crossed the German lines on their way out and slowed them on their return. The *Luftstreitkräfte* had only to wait over their own territory, between the Allied front line trenches and the returning RFC and RNAS aircraft, to intercept them. 'Let the customers come to the shop,' Manfred used to say. Apart from taking advantage of the headwind that handicapped the 'customers', there was another consideration: the German air service was outnumbered by the combined strength of the British and French in the air and could not afford to lose aeroplanes shot down onto enemy territory.

The official British history of the air war explains,

'Richthofen's task was to inflict the greatest damage with the minimum loss to his own service, and he knew that on any day suitable for flying great numbers of aeroplanes of the Royal Flying Corps would be over the German lines. He seldom had to seek combat. It was offered to him.'

He could make his choice whether to join action, to break off if that were prudent, or to avoid it in the first place.

It was now that Manfred became known as The Red Baron, *Le Petit Rouge* and *Le Diable Rouge* because he had the tail unit and wheels of his aeroplane painted red, then later the fuselage and ultimately the whole machine. Three reasons have been attributed to this: one, that he did so in order to be readily identifiable to the members of any formation he was leading; another, that both the Germans and the Allies had been experimenting with camouflage to make their aircraft more difficult to spot in the air and Manfred, impatient with the apparent futility of these experiments, figuratively cocked a snook at them; the third is that he hoped thus to frighten his opponents when they knew his identity. The first is reasonable, because, although British squadron and flight commanders flew coloured streamers from mainplane struts on either side of the fuselage, these could be seen only by those flying quite close. When a flight or squadron scattered it was often difficult to spot streamers if trying to re-formate. The second is not implausible, for he was impatient with any waste of time and had a quirky sense of humour beneath his sober demeanour; the third is convincing because it conforms to the inherent insolence of the aristocratic Prussian nature. He acknowledged none of these. He himself wrote disarmingly that he did so for no special reason, which is too naïf and out of character. It seems logical to conclude that he did so from flamboyant arrogance and to intimidate his enemies; although a really frightened pilot, particularly a novice, would have turned tail and deprived him of an easy kill – the last thing the Baron wanted.

Between patrols he lectured his pilots on tactics and air discipline, the importance of closing the range before opening fire and the uselessness of aerobatics. In this last, he was right only in part. Immelmann's half loop and diving turn off the top onto his opponent's tail was immaculately timed and executed, but in most instances a loop as an evasive measure presented a plan view of the evading aircraft and gave its pursuer a big target. Also, any aerobatic slowed the aircraft, which gave an attacker more time to close and aim. However, a good aerobatic pilot could in an extremity use his skill in evasion when a lesser artist could not. The least promising fighter pilot material in both world wars were instructors, whose polished flying became an unbreakable habit, whereas in a dogfight there was no time for delicate handling and accurate flying: the controls had to be treated roughly in evading attack and getting into position for a successful shot.

Many excellent aerobatic pilots were immensely successful in battle: Douglas Bader had been in the RAF aerobatic team and, ten years later, fighting with two artificial legs, shot down twenty-four enemy aircraft. Air Chief Marshal Sir Harry Broadhurst had led the tied-together aerobatic team as a squadron leader, was thirty-three when the war began and, three years later, when he became the RAF's youngest air vice-marshal, had a score of twelve victories. He had also led his squadron's air firing team of three, which won the RAF trophy – no small help

71

in action and confirmation of Manfred's insistence that good shooting was the most important factor.

In Manfred's day, there were few aerobatic experts and there was little time to spare for beginners to practice. In 1940, No. 13 Group of Fighter Command issued a booklet, *Forget-Me-Nots For Fighters*. The foreword explains,

'This book is the outcome of discussion amongst the Training Staff, on the best and simplest way to bring to the notice of new Fighter Pilots certain salient points in air fighting, which it is essential that they should master before taking their places as operational pilots in Fighter Squadrons'.

The guidance it gave would have been equally applicable in 1914–1918. Under the heading 'Evasive Tactics', the first item is,

'A barrelled aileron turn is very effective with fighters. An increasing rate of turn prevents the enemy getting his sights on you, and will usually give you a shot at him.'

A barrelled manoeuvre is an aerobatic. It might have saved Hawker's life. As for the increasing rate of turn – he must already have been flying as fast as practicable in a steep turn and, anyway, Manfred had much the faster aircraft.

Germany began to emulate France in forming élite squadrons. *Jasta* commanders with high scores could ask for outstandingly successful pilots to be posted to them. Some of those whom Manfred began to gather around him because they had already drawn his attention by achievement or showed promise of it were destined to command squadrons and win the Blue Max. Among them were three *Leutnants*, Karl Almenröder, Kurt Wolff, Emil Schäfer, and *Unteroffizier* Sebastian Festner.

His pilots soon became uneasy about the danger in their Commanding Officer's flying a machine whose paintwork was so easily distinguishable. They persuaded him to let them have their aeroplanes 'personalised' in the same way. He agreed to their having various components – tail units, struts, wheels, portions of the fuselage, etc – coloured according to each man's fancy. One chose black for the rudder, elevators and rear half of the fuselage, others' decorations were in all the colours of the spectrum plus one that was all pink. The RFC called them the Flying Circus. One of the many rumours in circulation for a while was that the one dubbed the Pink Lady was actually flown by a young woman.

To date, all Manfred's kills had been made when flying a Fokker E type or an Albatros D type. On 1 February he went up in a Halberstadt DII, which was a trifle slower in level flight than the Albatros but ninety seconds faster in reaching 10,000 ft, with Almenröder as his wing man. They caught a BE2d unawares and Manfred shot it down. Pilot and observer died of wounds. Manfred resumed flying the Albatros.

Early in February he had to go to Berlin to discuss the evident weakness in the Albatros III's lower mainplanes. Thence, he went home to show his mother his *Pour le Mérite*. She reproved him for making himself dangerously conspicuous to the enemy by the gaudy decoration of his aircraft, but this did not deter him from retaining its livery.

Winter weather inhibited activity, but on 14 February, returning to base from a visit to *Jasta* 2, he came upon an artillery spotting BE2d, shot it down, wounded the pilot and killed the observer. In the afternoon, leading a formation of six, he interrupted the work of two BE2cs. While his 'gentlemen', as he often referred to them, attended to one, he sent the other down. Its crew, who had as much chance as a celluloid cat in Hell, got off lightly: the pilot wounded, the observer unscathed. This easy success was his twenty-first. In March he notched up ten more, on the 4th (two), 6th, 9th, 11th, 17th (two), 21st, 24th, and 25th. The aeroplanes he sent down were, in order, BE2d, Sopwith 1½–Strutter, BE2e, DH2, BE2d, FE2b, BE2c, BE2f, Spad S7 (*La Société Pour l'Aviation et ses Dérivés*), and Nieuport 17. The fates of the crews were, respectively: pilot wounded, observer unhurt; both killed; both killed; single-seater, pilot killed; both killed; single-seater, pilot killed; both killed; single-seater, pilot wounded; and single-seater, pilot unhurt.

Most of these successes were against aircraft inferior to his. The FE2b was slower and less nimble. His victory over the

DH2 of Second Lieutenant A.J. Pearson MC of 29 Squadron proved the superiority of the Albatros DIII in all respects over the hitherto dominant British fighter. The one over the Spad, flown by Second Lieutenant R.P. Baker of 19 Squadron, was fairly creditable. To get the better of the Nieuport 17, capable of 135 mph and the fastest machine at the Front, flown by Second Lieutenant C.G. Gilbert of 29 Squadron, was meritorious. Nonetheless, one has to take into account that the Albatros's twin Spandaus gave him a vast advantage over all these single-gun machines.

During this orgy of bloodshed, Manfred had suffered a defeat on 9 March – not a fatal one, which would have saved the lives of many Britons. He was leading four others of his squadron at the apex of the usual V formation he favoured, at 3,000 metres altitude, when they intercepted nine FE8s of 40 Squadron on barrage patrol: unwieldy, lattice-tailed 69 mph affairs that were no longer fit to qualify as fighters. Even though they outnumbered the enemy, they were as much at a disadvantage as nine sheep being attacked by five wolves. They held formation, which gave them the best chance of some, at least, surviving. In the early moments of the fight, when Manfred had fired only ten shots, he was hit by tracer bullets from fifty yards, then: 'Suddenly I heard a tremendous bang and something hit my machine. I was thankful my aeroplane had been hit and not I. At the same instant there was the smell of petrol and I switched off the engine'. Spillage onto the hot engine could set the aircraft on fire. He dived away as a burning FE8 began to fall and, in constant fear that his own machine would ignite, landed in a meadow near a main road. A car arrived to take him back to camp and he was airborne again soon after. He compensated for his misadventure by hacking down Pearson's DH2.

Lothar was able to admire, and be inspired by, his elder brother's last six kills of the month immediately they occurred: on 10 March he had joined *Jasta* 11; not because Manfred asked for him, but because higher authority was awake to the publicity value of the two fraternal *Freiherren* operating together.

On 28 March Lothar had his first victory by forcing an FE2b down behind the German lines, its pilot dead, the observer wounded.

CHAPTER 10

THE LUFTSTREITKRÄFTE'S RECORD MONTH, APRIL 1917

In September 1916 there were seven Albatroses at the Front. Two months later there were seventy-eight. By January 1917 their numbers had increased to 270, by March to 305 and in May reached 434. They were flown by thirty-seven *Jastas*, each numbering fourteen at full strength. German fighter operations were served by the Flight Reporting Service, which received information about enemy aircraft approaching or over the German lines, from observation balloons, spotters in church steeples, and ground troops. This was passed to the aerodromes by telephone. Because they stayed behind their own lines, German fighters were able to climb high and up-sun while awaiting the arrival of the British or French.

The Battle of the Somme in July 1916 had launched that year's big offensive. The British infantry attacked after the manner of troops in past centuries, across open ground, rank after rank. By the end of the day they had suffered 60,000 casualties cut down by machine-guns, 20,000 of these killed and most of the others dead too because the Germans had taken few prisoners. And that was only the beginning of the three months' slaughter.

'Must do better next time', the obtuse British General Staff resolved; and so, for 1917, a titanic offensive on the Arras Front was planned, to open on 9 April and be followed a week later by a massive French assault in Champagne. The British were immediately successful and the next two months brought more gains in ground. It was the French who were to be the victims of poor generalship this time: in May there were widespread desertions and sixteen of their corps mutinied. It took a month to restore good order and discipline.

The Albatros DIII remained the best fighter at the Western Front despite the failure to identify and rectify the structural weakness that was causing frequent crashes. Aerodynamics had not yet led to the discovery of wing flutter. The DIII was a sesquiplane, i.e. its lower mainplane was shorter and narrower than the upper. This improved manoeuvrability and the pilot's downward view. But there was only one strut on each side, near the wingtips and V-shaped, with the apex exerting pressure on the lower wings' weakest point where, under excessive torsion, it broke.

In spite of this defect, fatal accidents were few in proportion to the number of sorties flown and the Albatros remained master of the air. British Flying Training Schools were having to send newly qualified pilots to the Front with so few flying hours that their life expectancy had shrunk to between eleven and twenty-one days. The French were in almost the same predicament. The Germans' training remained more rigorous than either, which meant that although their supply of new pilots was slower than the oppositions', the demand for replacements was lower because the new entries at the Front were more competent than Britain's or France's.

The fourth month of 1917 became known to the RFC as 'Bloody April'. The scale of air casualties can be judged from the fact that Manfred had twenty-one victories in twenty-eight days. His varied bag numbered seven BE2s, two Sopwith 1½-Strutters, five FE2s, an RE8, two Nieuport 17s, a Spad S7, two Bristol Fighters and a Sopwith Triplane. He killed eleven pilots and ten observers; wounded six pilots, of whom one died; and five observers, two of them fatally.

The new Bristol was the first of its type to be called a fighter. Hitherto the word 'scout' had been used and continued to be as a generic term. The Germans' nomenclature, *Jagd*, meant 'hunt', and the French had always called them *avions de chasse*, (hunting aeroplanes).

The F2b Bristol Fighter was a big, strong two-seater. A synchronised Vickers gun was mounted for'ard, while a Lewis gun on a Scarff ring covered 180 degrees astern. Over twenty-six feet long, with a wingspan of nearly forty feet, the 'Brisfit' had a top speed of 111 mph at 10,000 ft, to which it climbed in thirteen minutes. Despite its size it was highly manoeuvrable. No. 48 Squadron was the first to receive it. So desperate was the situation in France that the squadron was sent there in late March 1917 although only six had been delivered.

On 5 April the new scout went into action for the first time when Captain Leefe Robinson VC, a flight commander who had won his decoration by shooting down a zeppelin near London, led a formation of five. They were intercepted by five Albatros DIIIs led by Manfred. The RFC had not yet appreciated that the pilot's heavy machine-gun should be used for attack and the observer's light one primarily for defence. The 'Brisfits' therefore adopted the standard tactic of positioning themselves so as to give the observer the best possible field of fire. Consequently, Manfred shot two down, his companions took out another two and the fifth returned to base badly damaged. Henceforth 48 Squadron attacked head-on, as did all the other squadrons that flew this formidable aeroplane. It was so successful that it was ultimately regarded as the best all-round aeroplane of the war.

After that initial fight, Manfred described the Bristol Fighter with contempt to newspaper reporters – a great disservice to his fellow *Jagdflieger* – who, when they first encountered one or a formation of them, expected to make easy kills; until they learned that their star performer was not infallible.

He had been promoted to *Oberleutnant* on 22 March. On 7 April he went up another rank, to *Rittmeister*, captain of cavalry.

On 13 April he scored his first kill, an RE8, at 8.55 a.m., thus beating the record of forty-one victories set by Boelcke. He made a second kill (FE2b) at 11.45 a.m. and scored a third (FE2b) at 7.30 p.m.

The High Command, worried that he might be shot down, had told him to go on leave immediately he had surpassed Boelcke's score. He ignored the order: an extraordinary departure from the tradition of a family such as his and the discipline under which he had been brought up. He denied that he did so because amassing the biggest possible score took priority in all his actions. His attitude implied that he went on fighting because he felt physically and mentally fit to do so without a rest. In view of his lust for collecting trophies of the aeroplanes he shot down and the silver commemorative cups, this claim is not entirely convincing. Sheer pot-hunting, patriotism and a sense of duty that, he felt, justified his continuing to lead and fight seem the most likely motives.

On the 29th Manfred shot four down: a Spad S7 at 12.15 p.m., an FE2b at 4.45 p.m., a BE2e at 7.25 p.m and a Sopwith Triplane at 7.45 p.m.

About the BE2, his combat report was, 'With my brother, we attacked an artillery spotter at low altitude. After a short fight, my adversary's machine lost its wings. When it hit the ground it caught fire.' So, when they opened the attack together, it was 'we', but when the BE2's wings were shot off it was 'my'. Lothar shot the other one down, his tenth score.

Between them, Manfred and five others of his squadron had thirteen victories that day, of which Lothar was credited with two. It was perhaps this orgy of homicide that caused Manfred to admit, when writing about it, that he still had nightmares about the first Briton he had killed.

Reminiscing about the blood-soaked day, in his autobiography, he changed the subject as though it were distasteful, by waxing sentimental over his dog.

'Ever since Ostend he has accompanied me and grown in my affection. The little lapdog has become enormous. He is now a year old, but still a puppy. He plays billiards very well. Unfortunately many balls and table cloths have been ruined. He also has quite an instinct for hunting, about which my mechanics are pleased because he has brought them many nice hares. But he gets a good thrashing from me for it, as I am not pleased by this.'

Evidently affection did not over-rule cruelty; nor, since the dog continued to hunt hares, was it effective, therefore sheer brutality and a waste of time. The wholesale carnage he perpetrated on pig and various species of deer suggests that he had no love for animals. A possible inference is that, being short in stature, he compensated for a feeling of inferiority by dominating animals, so subjugating a huge dog was flattering to his ego. But to a Prussian, especially one who had been under military training from an age when most boys would be playing childish games, inflicting pain was not sadism, it was discipline.

As he flogged his pet, one wonders how he treated his horses and how much whipping he inflicted when schooling them over jumps and in a race. Before passing judgment, however, one should remember that he was a cavalryman, expecting to ride into the violence of battle. A well disciplined horse was essential to the effectiveness and survival of both horse and man.

He goes on to say that the dog had the bad habit of chasing aeroplanes when they made their take-off run and had once been clipped by a propeller, which cut half an ear off 'and a very beautiful propeller was ruined'; which insults one's intelligence.

The Sopwith Triplane that fell victim to him belonged to No. 8 Squadron of the Royal Naval Air Service. Since July 1916 these and Sopwith Pups, both single-seaters, had been operating on the northern sector of the front with conspicuous success. The crisis on the Somme Front had prompted the RFC to request naval assistance. A Sopwith Pup squadron, No 3 RNAS, and two Triplane squadrons, 8 and 10 RNAS, were detailed to provide it. The 'Tripehound', as the Navy called it, was armed with a synchronised Vickers, sometimes with two. All three of its mainplanes were narrow, which allowed the pilot an excellent all-round view. It was the most manouevrable aeroplane in the world and the fastest climber; in twelve minutes it could make 10,000 ft, where its top speed was 110 mph. Manfred considered it the best Allied fighter during the first eight months of 1917. No. 10 Squadron RNAS was commanded by a Canadian, Raymond Collishaw, the RFC and RNAS's third-highest-scoring pilot of the war, another with whom Manfred will be compared.

The new record-holder was ordered to go on leave at once, so, on 1 May he handed over command of the squadron to Lothar.

This was unfair to those who were senior to Lothar, had more flying hours and more kills; but it was excellent propaganda for the press to spread. The acting Commanding Officer was entrusted with leading a formation of twenty-two aircraft from his own *Jasta* and three others: while so doing he distinguished himself by shooting down a BE2g and a FE2d, which raised his total to sixteen.

One of *Jasta* 11's main interests just then was a rumour that the RFC was forming a special anti-Richthofen squadron to wipe it out. A news agency issued a story, which the newspapers printed in articles declaring that this mythical unit consisted of volunteers and the pilot who shot down the great man would be given the VC, £5,000 and promotion. Furthermore, a photographer with a ciné camera would be taken up to record the event. This balderdash discounted the fact that if it were true, then the RFC would not rely on volunteers; it would detail those most likely to out-fly and out-shoot the intended target.

Manfred, on leave, wrote to the widely-read *Vossische Zeitung*, tongue in cheek, to point out another obvious fatuity; supposing he shot down the cameraman? Here again, a sense of humour emerges through his shyness and reserve.

CHAPTER 11

SOME BRITISH ACES

Although not specifically formed for the purpose of removing Manfred von Richthofen from the scene, a new squadron, equipped with a new fighter, had become operational at the Western Front on 22 April. Emulating the established French and German system of forming élite flying units, its pilots were carefully chosen for their outstanding skill, aggression and experience to constitute the first RFC squadron in which they were all well above average. Captain Albert Ball DSO, MC, vanquisher of 31 hostiles, was the best-known of them and the only one with a DSO. Three of the others had an MC. The British were less liberal with decorations than the French, Germans and Italians. The squadron's number was 56 and its aeroplane the new Royal Aircraft Factory SE5, whose 150 hp Hispano-Suiza engine gave it a top speed of 120 mph at 6,500 ft and a ceiling of 18,000 ft. There was a synchronised Vickers in front of the cockpit and a Lewis on the upper mainplane. Its endurance of two and a half hours was an hour better than the Albatros's, which meant that its pilots had ample time to stay at great altitude, waiting to pounce on the enemy.

Lieutenant-General Sir David Henderson had been the first Commander-in-Chief of the RFC when it was formed on 13 May 1912. He served in France from the outbreak of war until 20 December 1914, when he returned to its Headquarters in London. In the first week of April 1917 he wrote, for the information of the Chief of the Imperial General Staff, a summary of events that had affected the RFC at the Western Front. Its rambling contents, with a few verbosities omitted, were as follows.

'The increased number of casualties in the field are due to several causes. In the first place, the retirement of the Germans over a large section of the front necessitated a great amount of long-distance reconnaissance and photography. This is always dangerous work, and specially dangerous in this case because of the special efforts made by the Germans to stop it.

'Probably, in view of this retirement, the Germans had concentrated a very large proportion of their available forces in front of the British. There has not been so much fighting in the French part of the line, and this may also be due to the fact that the French Air Service, with regard to their front, was very incomplete, so much so that a considerable portion of the German line in front of the French had to be photographed by the British Flying Corps.

'There is no doubt that the Germans have produced within the last few months, a considerable number of fast single-seat Scouts, of which the best is the Albatross [sic]. The aeroplanes which we have on our front which are equal to, or better than the Albatross Scout, are two French types – the Spad and the Nieuport – and the English Sopwith Triplane. Next to them, and still able to hold their own, are the small Sopwiths and the Martinsyde squadrons. Our first-class two-seater machines capable of begin used for offensive fighting, are the de Havilland 4 and the Bristol Fighter: there are at the moment 1 squadron of each. The FE2d, with the Rolls-Royce engine, is a two-seater Fighter, which will not be outclassed for some time: of these there are 3 squadrons. The machines principally used for reconnaissance are Sopwith 1½-strutters: of these there are three squadrons. A squadron of SE5 single-seat Fighters, which is believed to be superior to any German machine, is due to leave England this week.

'The delay in producing large numbers of these fighting machines is due almost entirely to the delays in engine production. We are only now beginning to get British-made engines equal to those which the Germans had for the last eighteen months, with the exception of the Rolls-Royce engine, of which the supply has always been limited. The high powered British engines, however, have now reached the production stage, and the quantities delivered are expected to increase week by week, which will enable us to provide for the Expeditionary Force first-class fighting machines in good quantities.

'In addition to long reconnaissance, a very large amount of Artillery observation work is going on, much more in our Army than in either the French or the German. This certainly adds to our casualty list without inflicting on the enemy proportionate losses in the air. It does, however, enable our Artillery to inflict much more serious losses on the German forces on the ground, and this must be taken into account in considering whether we get sufficient value for the casualties we suffer.

'With regard to the losses inflicted on the Germans, the announcements which are made in the official communiqués do not show their full extent; so much of the fighting takes place on the German side of the lines that very often there is no information whatsover about the actions of our aeroplanes which are reported missing, but it is known that frequently in these unseen fights serious losses are inflicted on the Germans.

'It was noticeable last year that up to the beginning of June there was no marked superiority in the air on either side, and that the losses on each side appeared to be about equal. After that date, in the continuous good weather, our superiority became more and more marked, but our losses did not diminish to any great extent, for the reason that our superiority on the battlefield was only sustained by continuous fighting at a distance behind the German lines.

'If we would consent to adopt the same policy as the Germans, there is no doubt that our casualties in the air could be diminished. Hitherto, when the German has found himself inferior, he has given up reconnaissance entirely, and has confined himself to defensive fighting on his own ground, but if we were to follow these tactics the effect on the Army generally would be most serious. Such a policy at this period would be disastrous. The casualties must be faced.'

The basic fact is clearly that the RFC was the servant of the ground forces and the reconnaissance and artillery spotting aircraft were as important as the fighters. The woolly sentence about Sopwith Pups and Martinsydes 'holding their own' is meaningless. The RFC fighter pilots would not have agreed that the FE2d would not be outclassed for some time – it had been by the Fokker E series and was now by the Albatros D. One adroit omission is an explanation of why, after such horrific losses by BE2 squadrons, these archaic aeroplanes were still being built and why the death-trap RE8 had not been scrapped.

No. 56 Squadron was expected to reverse the situation at the front in the manner of the US 7th Cavalry rescuing a wagon train beset by Apaches.

It had formed at London Colney in the last week of February, under the command of Major R.G. Blomfield, who had a sense of style that consorted well with the panache displayed by his young comrades, most of whom were aged between nineteen and twenty-one. At the beginning of the squadron's six weeks at its home base, familiarising itself with the SE5, practising gunnery, aerobatics, formation flying and all the other necessary drills, he set about forming a band to play every evening in the officers' mess. Conscription had been imposed, so he visited orchestras in London restaurants and hotels to obtain the names of their musicians who had been called up, then pulled strings to have them posted to the squadron. He also had himself driven in a lorry with air mechanics (they were not called aircraftmen until the RAF was formed on 1 April 1918) of various trades aboard, to nearby RFC stations to swap them for men of the same trades who could play an instrument.

His high-spirited pilots invented a new approach to landing. The hangars had sloping roofs, so they touched down with their wheels on the side facing the airfield and rolled down it before settling on the grass: a feat that required great lightness of hand.

They were adversely critical of their new machine, Ball most scathingly so, and were allowed to have modifications done.

84

The first general objection was to the windscreen, which they complained became blurred by scratches and oil. The Lewis gun, for which all the ammunition was in double drums, was very difficult to load because of the wind resistance. Ball, writing home, described the SE5 as 'a dud whose speed is only about half a Nieuport's,' (in fact it was 15 mph faster) and 'a rotten machine', but he was 'making the best of a bad job'. This must have been more than a trifle worrying for his parents. Renowned for the importance he put on speed, he had the Lewis gun removed to save weight. All the pilots had the windscreen lowered or removed to decrease wind resistance.

Familiarisation changed these hasty assessments. Firing practice showed it to be a steady gun platform, highly responsive to the controls and a joy to aerobat.

It went into action for the first time on 22 April, the day on which Manfred scored his forty-sixth victory – over an FE2b. On 56 Squadrons's first patrol they shot down four Albatroses, of which Ball bagged one.

The pace thereafter was hectic. On 29 April, Ball's letter to his parents read,

'I am so very fagged. April 26 evening I attacked four lots of Huns with fire. Brought two down and had to get back without ammunition when dark. Had four fights and got one Hun. In the end all my controls were shot away. But I got back. Simply must close for I am fagged'.

That he should disturb their peace of mind with such frankness indicates how close he was to them and how ungregarious.

On 3 May his letter home, expressed in terms more suggestive of a 15-year-old schoolboy than a man of twenty, said,

'It is quite impossible, but I am doing all I can. My total up to last night was 38. I got two last night. Oh! It was a topping fight. A few days ago all my controls were shot away on my SE5. But I got the Hun that did it. It is all troubles. I am feeling very old just now.'

On 5 May:

'Dearest Dad, Have just come off patrol and made my total 42. I attacked two Albatros scouts and crashed them, killing the pilots. In the end I was brought down but am quite OK. Oh! It was a good fight and the Huns were fine sports. One tried to ram me after he was hit, and only missed by inches. Am indeed looked after by God, but oh, I do get tired of always living to kill, and am really beginning to feel like a murderer. Shall be so pleased when I am finished.'

On the same day, talking to Atkins, a pilot in the squadron, he said, 'Trenchard (RFC Commander in France) says I can go home when I have got fifty. But I shall never go home.' This can, with equal logic, be taken either as a premonition of death in action or as a resolution to remain at the front until ordered home.

On the evening of 7 May, eleven aircraft of 56 Squadron took off for the day's last patrol. At 18,000 ft, behind the German lines, they saw six Albatroses 3,000 ft below. About a mile away, two Albatros squadrons in full strength of twelve, one of them led by Lothar von Richthofen, also came in sight. Immediately they saw the first six, the SE5s attacked. The two other enemy formations, seeing them dive, went after them. Forty-one fighters scattered in a dogfight in which six British machines were shot down, Ball's among them. It has always been suspected that the six Albatroses were bait in a trap. Ball's score was then forty-three, currently the RFC's and RNAS's highest. It is not known how many he shot down in that engagement, but, from the description of the battle given by a pilot of *Jasta* 11, it is assumed that it was Ball who wounded Lothar and sent him down to a forced landing. On 3 June 1917 his award of the VC was gazetted. Although ten other British and Commonwealth pilots exceeded his number of victories by the end of the war, he is generally held to be one of its five greatest fighter pilots.

Lothar insistently claimed to have shot Ball down. A Vickers gun and other items from the wreckage were scavenged on his behalf for display at his family home. His fellow pilots thought it most likely that in the confusion of so many aeroplanes milling about, a bullet from any of the Germans could have hit Ball.

Years afterwards, investigation indicated that it was possibly a machine-gunner on a church tower who had done so.

One of the pleasantest men among the most highly esteemed fighter pilots was James McCudden, who, aged eighteen, joined No. 3 Squadron in 1913 as a mechanic and accompanied it to France in August 1914. A pilot of that squadron, with one from No. 4, flew the RFC's first sorties of the war, on 19 August. None of the squadrons was equipped entirely with the same type of aeroplane: all had various mixtures of Farman F20. Martinsyde and BE2. The category of 'observer' had not yet been created. When a BE2 was sent on reconnaissance the occupant of the rear cockpit was either a pilot or one of the ground crew. McCudden often volunteered for the duty, armed with a rifle. On 20 November he was promoted to corporal. In July 1915 he applied for pilot training but this was not approved, as he was doing such good work as mechanic and unqualified observer. Observer was an official category by then, symbolised by a badge, an O with a wing sprouting from it, embroidered in white and worn on the left breast. In compensation he was given the trainee observers' test, which he passed. In December, on a sortie in a Morane two-seater, which had a Lewis gun, he had his first fight when a Fokker EI made three attacks, all of which he drove off.

By January 1916 he was a flight sergeant and was at last sent home on a pilot's course. On 8 July he returned to the Western Front as a flight sergeant pilot in 20 Squadron, was posted to 29 Squadron the following month and shot down his first enemy aircraft on 6 September. On 1 January 1917, already holding the Military Medal and *Croix de Guerre*, he was commissioned. On the 26th he shot down another and on 23 February, with his score at five, went home as an instructor.

For nearly five months he taught new pilots combat technique. Flying a Sopwith Pup, in which he performed his first roll, he visited fighter airfields to instruct pilots stationed there. One day, on landing at Croydon, he learned that the first big daylight raid against London had just been made. None of the enemy aircraft had been shot down by the ninety-three Home Defence fighters and the anti-aircraft guns that had tried to do so. Fifteen minutes later he arrived at his own airfield, in Kent, to pursue them in a Pup armed with a Lewis gun. At 15,000 ft they came in sight, fourteen twin-engined Gothas whose crews had left 162 dead and many wounded behind them. He could not overtake them. Twenty miles out from the Essex coast he emptied three drums at the nearest one from 500 ft range and smelled the incendiary bullets with which they returned his fire.

He expressed the revulsion and anger he felt on seeing the enemy in British air space:

'How insolent these damned Boches did look, absolutely lording it in the sky above England. I was absolutely furious to think that the Huns should come over and bomb London and have it practically their own way. I simply hated the Hun more than ever.'

Twenty-three years later Douglas Bader vented the same feeling. Leading his fighters into battle for the first time against

'a horde of German bombers (plus fighter escort) flying at 17,000 ft in perfect formation. There were nine of us Hurricanes. Suddenly I was angry. "Who the hell do these Huns think they are, flying like this in their bloody bombers covered with iron crosses and swastikas over our country?"'

Some fighter squadrons were withdrawn from France to defend Britain, among them No. 56. McCudden knew several of the pilots, went to visit them and reported that they had a wonderful spirit which was entirely different from any other squadron's.

On 7 July he did a three-week refresher course, then was detached to 66 Squadron in France for up-to-date combat experience, after which he resumed instructing in England. No. 56 Squadron had returned to the Western Front and invited him to dinner while he was with 66; the orchestra played. He asked Major Blomfield if he would have him in the squadron and Blomfield promised to ask for him.

On 15 August, he was back in France again, as a flight commander in 56.

All aspects of aircrew training had vastly improved in the past three years and, although there were still outstanding pilots on both sides who preferred to go hunting alone whenever they could, McCudden among them, air fighting was now done mostly by small fighter formations of five or six. Line astern, line abreast, echelon, arrowhead were all favoured, depending on the purpose. Arrowhead, known as Vic (the signalling term for V), with two machines on each side of the leader, and a diamond of four with a weaver astern to give warning of an attack from that direction, were the most used.

On 18 August McCudden shot down an Albatros, another the next day, two on the 21st and a Halberstadt DFW two-seater bomber on the 22nd. In the evenings, at about 7 p.m. there would be eight British fighter formations aloft, among various other types. The Germans usually sent up about the same number of fighters. According to him, 'The evenings were wonderful, as the fighting was very fierce and well contested.'

McCudden ended 1917 with a fine metaphorical firework display. On 23 December he shot four down. On the 28th three fast LVG two-seater, two-gun bombers in twenty minutes. He surprised the first LVG by a stern attack, about which he confessed some shame. 'I hate to shoot a Hun down without his seeing me, for although this is in accordance with my doctrine, it is against what little sporting instincts I have left'. He sent the next one down from a distance of 400 yards. He knew his estimation of the range would be disputed, but was adamant that his judgment was accurate.

As a former mechanic, he was observant of the varying performance by aircraft of the same type. He was irritated when a Rumpler C IV, which had a 260 hp Mercedes engine, out-climbed him when he was at between 15,000 and 16,000 ft. The SE5a with a 200 hp Wolseley engine came into service shortly after and he had his fitted with high compression pistons. 'I was very keen to see the Rumpler pilots' hair stand on end as I climbed past them like a helicopter'. It is interesting that although no helicopters had yet been built, he knew of them: perhaps from reading about Leonardo da Vinci's design.

His life as a regular in the ranks had taught him intolerance of any form of slackness. On 24 January 1918 he shot down a DFW that was artillery spotting at 12,000 ft.

'This crew deserved to die, because they had no notion whatever of how to defend themselves, which showed that during their training they had been slack and lazy. They probably liked going to Berlin too often instead of sticking to their training and learning as much as they could. I had no sympathy for those fellows.'

He could be admiring too. One day, leading a patrol, he saw a DFW at 4,000 ft, below the clouds. He signalled to the other four in his formation to stay above cloud in case the DFW escaped him, while he went down to attack it. He found that his Vickers had jammed, so had to rely on his Lewis. Losing height during the five-minute fight, he broke off at 500 ft. 'The Hun was too good for me and had shot me about a lot. Had I persisted he certainly would have got me, for there was not a trick he didn't know, and so I gave that liver-coloured DFW best.'

Well known was an Albatros with a green tail that he had often seen in action, its pilot showing consummate skill at shooting down British aeroplanes, including some from 56 Squadron. One day he and his flight met a flight led by Green-tail, at whom they shot and set his aircraft alight. The German pilot fell, or jumped, out.

'He was hurtling to destruction faster than his machine. I now flew on to the next Albatros and shot him down at once. I must say the pilot of the green-tailed Albatros must have been a very fine fellow. I had many times admired his fighting qualities. I only hope it was my first bullet that killed him.'

On 16 February he increased his score by four. A fortnight later he was posted home, promoted to major, and in April was awarded the Victoria Cross. His death came in the last way that anyone would have expected. On 8 July he was given command of 60 Squadron. He had barely become airborne on his way to his new station, when his engine cut. Like an unwise novice he did the strictly prohibited thing: turned back to try to land on the airfield. Predictably, the aeroplane stalled into the ground and he was killed. His victories totalled 57.

The highest-scoring pilot in the British and Commonwealth air forces was Major Edward 'Mick' Mannock VC, DSO, MC, who shot down 73 enemy aircraft. He was highly eccentric, neurotic and extreme in his views, political, social and of the enemy. His career in the RFC is all the more extraordinary because he joined it so late in the war. In origin as well as by nature he differed greatly from the well-born Hawker and the lower-middle-class McCudden. The latter, who had joined as a regular and served for nearly four years in the ranks, during which he had exerted the authority of a senior NCO and flown with many commissioned pilots, had no prejudice against officers. In those class-conscious days he had experienced the formalities of the sergeants' and warrant officers' mess. He also had ample time to observe the way in which commissioned officers behaved and the shibboleths of social convention to which they were brought up. When he was commissioned he felt immediately at home in the officers' mess. In Mannock, detestation of class distinction and privilege was as relentless as his hatred for Germans. His position of top-scorer gives him the undeniable right to be ranked, with Hawker and Ball, among Britain and the Commonwealth's five greatest, whatever personal qualities he lacked in comparison with them.

Born on 24 May 1887, he was the son of a hard-drinking Irish corporal in the Royal Scots Greys and later 5th Dragoon Guards. That the senior Mannock enlisted as a regular, evidently owing to a poor education, is surprising – his father had been editor of a Fleet Street newspaper. Even more surprising is that Mick passed the RFC air crew medical, let alone that he was so brilliant with a machine-gun: in India, when aged ten, he was blinded for two weeks by a dust-borne amoebic infection that left him with corneal damage to his left eye, which permanently impaired his vision. His father taunted him about his poor sight and told him he would 'never be a real man', by which he meant a soldier. Mick had also contracted malaria. He showed his courage at a very early age: whenever his brutal, drunken father threatened to thrash him and approached with raised fist, he did not retreat but confronted him and the bully always desisted. He used to say this taught him that even a false show of fearlessness would discourage an attacker.

After the Boer War, Corporal Mannock had served his time and became unemployed. When he had squandered the small sum that he and his wife had managed to save, he deserted her and their three children. Mick got a job with a greengrocer at two shillings and sixpence (12½p) a week and next with a barber at twice the wage. In 1903 he joined the National Telephone Company as a clerk. In 1906 he transferred to the Engineering Department as a labourer, which entailed climbing telegraph poles to do repairs in all weathers. To fill his spare time profitably by earning a few shillings for attending parades, he joined a Territorial unit of the Royal Army Medical Corps, in which he rose to sergeant. Although, as a result of his abominable child-hood and penurious young manhood he had become a ranting, proselytising socialist, he was a great patriot who would verbally challenge anyone who denigrated Britain or its monarchy.

In February 1914 he went to Turkey, hoping to find work on a project there for building telephone exchanges and laying cables; which he did, as a supervisor. When war was declared in August, Germany began negotiating an alliance with the Turks but the work continued. In November the British Ambassador was recalled and all British residents were made prisoners of war. The Turks intended to repatriate them but the Germans objected. Mannock and others tried unsuccessfully to escape and eventually a Turk who used to work for him and was visiting him in the concentration camp, agreed to cut the wire so that he could break out every night to buy food for himself and his fellow prisoners; until he was caught and put in solitary confinement. Through American intervention, the British were sent home in January 1915, Mannock ill with a bout of malaria.

The hatred he already had for Germans was exacerbated when he heard about their use of gas and other – some apocryphal – atrocities. According to his best friend, 'His blood ran hot. Even his waxy complexion could not conceal it. His face reddened and I saw his knuckles growing white as he clenched and unclenched his fists in a growing fury.'

In July 1915 he rejoined the RAMC and was soon made a sergeant again. Hoping to arouse an aggressive spirit in his comrades – surely inappropriate in medical orderlies – he used to harangue them with tales about Turkish and German cruelty. Demonstrating how to treat wounded, his accompanying imaginative descriptions of front line dressing stations were lurid with gruesome detail about mud, filth, mangled limbs and enemy shelling. The thought of perhaps having to attend to enemy wounded was so repellent that he asked for a transfer to the Royal Engineers.

Eventually, he managed to be accepted by the RFC for pilot training and on 7 April 1917 arrived at the Western Front as a member of 40 Squadron, which flew the Nieuport 17. When the Commanding Officer, Major Tilney, described by another member of the squadron as 'a rather florid-faced youth', took him to the officers' mess to introduce him, the impression he made was obnoxious. Shyness and social clumsiness coupled with an eagerness to learn in detail about his new environment precipitated him into insensitive and seemingly impertinent questioning. Most of the pilots had just returned from a patrol on which the most popular member of the mess, Lieutenant Pell, had been killed, Mannock's bumptiousness in asking each of them how many 'Huns' he had 'fanned down', which was very bad form, guaranteed his instant unpopularity. Worse was to come. His new colleagues had begun their flying careers on pusher types. When training, he had acquired more hours on tractors than any of them yet had. Bolstered by this, he now launched into a dissertation on air fighting and, in general, expressed views on the war that they felt he was not qualified to. When they sat down to lunch, he took the chair that Pell had usually occupied; this, he did not know, but it unfairly added to the offence he had given.

A pilot who recalled Mannock's bombastic, conceited-seeming behaviour, said,

> 'Apart from that he was different. He seemed too cocky for his experience, which was nil. New men usually took their time and listened to the more experienced hands. He was the complete opposite and offered ideas about everything; even the role of scout pilots and what was wrong with aeroplanes. He seemed a boorish know-all.'

Until he was sent on his first operational sortie, Mannock practised firing at a ground target from different angles and at various speeds. For this, he was ridiculed behind his back, but perhaps many of those who made fun of it would have lived longer if, as Ball and McCudden did, they also had striven to perfect their gunnery in this way. Obtusely, he bragged about his accuracy and announced that if he closed to within twenty yards of an enemy machine he would bring it down. When he made his first combat flight he was so tense with nerves that he kept losing formation, a mishandling of throttle and controls that persisted. Because of his slow reactions he was the last to enter a fight and began to be suspected of cowardice. Ignored or treated contemptuously, he tried to make himself accepted by cutting into conversations. His worst breach of mess etiquette was an insistent introduction of politics as a topic so that he could press his left-wing views.

On 1 May his flight escorted four Sopwith 1½-strutters on a bombing raid against the Douai aerodrome, *Jasta* 11's base. When he tried to fire a short testing burst – 'clearing' the gun – it jammed. Thinking that if he turned back he would be accused of cowardice, he flew on. Hearing machine-gunfire behind, he turned about and saw a yellow and green Albatros diving at him. In retrospect he said that he heard a strange noise and realised that he was screaming with anger, which helped his nerves. Altogether, his conduct on the ground and in the air would seem to have made him a more suitable candidate for psychiatric treatment than for the distinction of becoming the RFC's top ace.

On 7 May, the day of Ball's death, he was on a balloon-busting operation for the first time. Major Tilney had perfected a new method of attacking observation balloons and Mannock was one of the five pilots whom his flight commander, Captain Nixon, led across the enemy lines to put it into execution. Hitherto, when the enemy saw a raid on balloons approaching at high altitude, there was often time to winch the balloons down. Tilney's innovation was an approach hugging the ground at fifteen feet, although this meant passing through successive belts of ground fire. This time, five Albatroses with Lothar von Richthofen in the lead were waiting for the Nieuports. Nixon indicated a target for each of his pilots. Mannock's was at one end of the line of widely-spaced balloons. He destroyed it with one long burst. Nixon had seen the Albatros and peeled off to try to protect his comrades. This drew the enemy's attention and while four Albatroses distracted his, Lothar shot him down.

On 9 May Mannock was No. 2 in a pair behind the enemy lines at 16,000 ft when his leader turned back with engine trouble. Mannock saw three Albatroses and turned to take them on. His gun jammed, so he spun out but they followed him down. He kept coming out of his spin to fly towards the British lines until, when he was safely behind them at 6,000 ft, the enemy retired. He freed the jam and returned to enemy territory, but found no trade. He was detailed to patrol again in the afternoon, but Tilney saw how tired and dejected he was so sent him as passenger in a 16 Squadron RE8 to fetch a new Nieuport from St Omer. On the way, Mannock experienced a change of attitude. Brooding on the fear he felt when in enemy air space or in a fight, he realised that he must overcome it and apply himself entirely to learning his trade. Thenceforward he was relaxed and more cheerful and the other pilots began to show friendliness.

Still he did not score a victory. To encourage him by a show of confidence, Tilney often let him lead a patrol; which increased his self-assurance but did not improve his fortunes. On 25 May he led his flight in an attack on two artillery spotters, of which one proved too high to catch up with. He fired thirty rounds into the other one's cockpit. It nosed down slightly but flew on. He was sure he had killed the pilot and the aeroplane was continuing its way as it had been trimmed to fly. None of his companions had seen him hit it, so he said nothing about it, for fear of disbelief.

On 7 June he shot down an Albatros, which was confirmed. Five days later he was coming in to land when he felt intense pain in his right eye. He fainted when the squadron Medical Officer attended to it. In hospital, under anaesthetic, a surgeon removed a large piece of grit and a sliver of metal. Another five days passed without scoring and he went home on fourteen days' leave, found that his mother had become an alcoholic, and returned gladly to France.

On 12 July he brought a DFW two-seater down and, when he went to see the wreckage, was sickened by the sight of two dead men and, in the rear cockpit, a small black dog, all three badly mashed up by his gunfire. He sent down two more two-seaters next day, which ended any suspicions about his willingness to fight.

A newcomer to the squadron at about that time, Second Lieutenant McLanachan, saw him in a different light from the one in which he had been regarded when he joined No. 40. On the afternoon of his arrival, McLanachan was surprised to see a tennis court, where a game of doubles was being played, watched by officers in deck chairs. He asked where the CO was. Someone pointed him out, on the court, 'but you'd better not disturb him until the set's finished'. Noticing the new arrival, Tilney called, 'You the new pilot? See you when we've won this set'.

When Tilney sat down to talk to McLanachan, he asked if he, straight from a training course, had ever Flown a Nieuport and was surprised when he answered 'Yes'.

'Here's a fellow from No. 1 Reserve Squadron, who's already flown a Nieuport. Let's see what he can do,' Tilney said.

No. 1 RS was commanded by the famous Smith Barry, who had revolutionised flying training.

McLanachan had climbed only to 1,000 ft when his engine cut. To everyone's astonishment, he performed one of No. 1 RS's favourite 'stunts', a spinning nosedive; which, everywhere but in Smith Barry's unit, was regarded as certain death. McLanachan landed safely and waited for the mechanics to re-start the engine.

Instead, he was told that the Major wished to speak to him. When he approached, Tilney turned and walked away. Hastening after him, McLanachan began to apologise but was cut short with a dismissive gesture.

'A tall, weather-beaten pilot was laughing', McLanachan recalled. It was Mannock, who explained that all the onlookers thought he was going to crash and kill himself. 'We don't like watching fellows kill themselves, and Tilney looked away when he thought you were finished.'

McLanachan feared that he would be sent back for further training, but Mannock told him, 'If you can handle a machine like that, we want you in this squadron.'

He was ever grateful for Mannock's friendly and encouraging words, which showed him that Mannock 'had a somewhat puckish sense of humour'.

A few days later, on patrol with two others, McLanachan saw a Nieuport catch fire, flown by a popular nineteen-year-old. The spectacle revolted him and the third pilot. The Germans were using incendiary bullets, which were forbidden by the Geneva Convention, except against balloons. For his next sortie, McLanachan told his mechanic to arm his gun with incendiaries, but the mechanic refused because he would also incur a court martial. McLanachan set about loading his gun himself, but Mannock talked him out of it: 'They've never fired anything at me but incendiary. Could you coolly fire that muck into a fellow creature; or worse still, into his petrol tank, knowing what it must mean?'

By the end of the year Mannock was a captain commanding a flight, had shot down six hostiles and been awarded an MC and bar. He was posted to an instructing appointment in England. During the past nine months he had thought deeply about tactics and become a fine leader. He realised that air fighting had developed to a point where team work and tactical judgment were most important. Like others before him, he constantly repeated his own essential rules. Boelcke's and Manfred von Richthofen's were long, as, twenty-three years later, were Malan's. Mannock's were brief: 'always above, seldom on the same level, never underneath'.

In March 1918 Mannock went as a flight commander to an SE5A squadron, No 74, which was working up for the front. He had developed an extrovert dimension to his naturally introspective character: he was the life and soul of the mess, where he led his brother officers in song and, with a pair of drumsticks, beat the rhythm on a tympani consisting of cans, tankards, pots, pans and glasses tied to the back of a chair.

The squadron arrived in France on 1 April 1918, the date on which the Royal Air Force was born. Mannock's analytical mind prepared for every patrol he led and, prompted as much by his natural garrulousness as by his innate thoroughness, he briefed his pilots carefully before taking off and de-briefed them with equal loquacity when they landed back. He was such a kind man that one of his many concerns was about the effect on his inexperienced pilots of seeing aeroplanes shot down in flames. In those days of easy conflagration and no parachutes, this sight could unnerve a hypersensitive youth. He himself was still highly vulnerable to the trauma that such spectacles could cause; and when one of the SE5As crashed on the airfield and he smelled burning flesh, it brought him close to a breakdown.

One evening, after Cairns, a particular friend of his, had been killed, he made sure that the mess gave him a 'good send-off'. After the customary riotous games had been played, he made a speech, probably his shortest ever: 'To Captain Cairns and the last dead Hun. Sod the Huns.'

In June, by which time the DSO and bar had been added to his medals, he went on leave, during which a second bar to his DSO, his promotion to major and his appointment to command 85 Squadron were gazetted. At his farewell to 74 Squadron, he wept. It was such habitual revelation of his feelings that endeared him to his comrades.

His new squadron also flew SE5As. He found its morale low. His predecessor, Major Billy Bishop VC, DSO, MC, who was the current RAF top-scorer with 72 kills, was a keen loner and unenthusiastic leader. Mannock interviewed all the pilots and got rid of those about whom he did not feel confident. There were three Americans, Elliott White Springs, Larry Callaghan and John Grider, all of whom he kept.

The first time he took the squadron up, he chose three men to go with him as decoys. The other flights followed at different altitudes. At 8.20 am he sighted ten Fokker DVIIs approaching. These had a best speed of 124 mph, the SE5As' was 130. Mannock and his decoys dived, followed by the enemy. On his signal, five of the flight next above his dived on the pursuers; then the third dived and took the enemy further by surprise. The battle began at 16,000 ft and finished at 2,000. Number 85 Squadron had no losses, the Germans lost five, of which Mannock brought down two.

Soon after this he heard about McCudden's death and became depressed, full of forebodings about his own. When he equalled Bishop's total and a friend he had invited to lunch told him, 'There'll be a red carpet reception for you after the war', he replied 'There won't be any after the war for me'. Later, when his friend talked about a flamer he had shot down, Mannock asked, 'Did you hear the swine screaming? When it comes, don't forget to blow your brains out'. Many pilots wore a revolver for that purpose, rather than be roasted to death.

He was shot down after one more victory and awarded a posthumous VC. His reputation stands above all his contemporaries of any nation in the First World War as what Air Vice-Marshal J.E. 'Johnny' Johnson, the RAF's second-highest scorer in the Second World War, describes as 'the master of team fighting, gunnery, ambush and decoy'.

Where Manfred von Richthofen stands in the scale by which these accomplishments are measured will be established when all the evidence in both great wars has been considered.

Chapter 12

French, Italian and American Aces 1914–1918

T he French air force had many gifted fighter pilots, few of whom were outstanding tacticians or formation leaders. The national attitude to life is selfish, so it is logical that there was an abundance of young airmen who wished to make their best possible demonstration of patriotism, and to distinguish themselves most, by individual achievement rather than as part of a team.

Italians, although by nature inclined towards flamboyance, and bravura performances that excite their own and their admirers' emotions, had many fine bomber, as well as expert fighter, pilots. That the number of the latter was much fewer than in the British and French air services is owed to the smallness of Italy's indigenous aircraft industry, late entry into the war on 24 May 1915 and the small size of their enemy, the Austro-Hungarian Air Service.

For those who are unfamiliar with it, *l'Armée de l'Air* system for identifying its squadrons is confusing. The squadron number was preceded by the initial letter of the make of aircraft it flew: thus, No. 3, *Les Cigognes*, flying the Morane Saulnier, began its existence as MS3; when it had the Nieuport it became N3 and when it flew the Spad it changed to S3.

Public adulation of the air services, above all, of fighter pilots, which surpassed admiration for the army and navy, was common to all the combatant countries. In France this reached its apogee.

In the early years of military aviation it was soon recognised that, whatever their nationality, pilots tended to be resentful of any discipline not directly connected to flying, were adventurous and usually had a streak of wildness in them. Whatever constraints they were under during their early training, this was relaxed when they qualified. In the peacetime RFC, for instance, it was common practice for officers to land in the grounds of friends' country houses on a weekend visit.

The French service appears to have been the most lenient. A pilot sergeant wrote, about his kith,

> 'Once called to the firing line he is treated on the same footing as an officer, whatever his rank'. Brindejonc de Moulins informed his parents, 'Apart from the guns which one clearly hears rumbling, this is the veritable country life lived in a *château* and, my word, I should greatly enjoy myself here. Yesterday I ate wild duck. Today it will be partridge for lunch and probably pheasant for dinner. What admiration we should have for the men in the trenches.'

Single-seater French pilots protested against any measure that tended to withdraw privileges from them and reduce them to the level of those in other combatant arms. To mitigate the ground troops' supposed jealousy, it was proposed to increase from five to ten the number of aerial victories needed to qualify for mention in despatches. Jacques Mortane, editor of *La Guerre Aérienne*, complained on the pilots' behalf in his columns and clinched the matter by denying that other arms envied them: 'Our troops are more tolerant, thank God!'

The most reckless and colourful of the brotherhood was Charles Nungesser, aged twenty-two when he enlisted in the hussars at the outbreak of war, with the intention of transferring to the air force. He was sent to the front on 20 August 1914. On 3 September he was given the *Médaille Militaire*. In January 1915 he transferred to the air service, qualified as a pilot on 2 March and on 8 April joined a Voisin bomber squadron, V106. On 22 April he was promoted to Sergeant for distinguishing himself in action. By 15 May he had flown fifty-three day and night sorties and was made a Warrant Officer. On 31 July he shot down an enemy fighter. In November he was posted to a fighter squadron,

N65, with which he spent the rest of his career. He celebrated the event by weaving among the chimneypots of Nancy, looped over the main square and flew along the main street at 30 ft. For this he was put in open arrest. Breaking arrest by ostensibly taking an aircraft on an air test, he shot down another German machine.

He had many crashes and suffered numerous injuries that would have discouraged the average man. He summed up his hare-brained method of fighting: 'Before firing my gun I shut my eyes. When I reopen them, sometimes the *boche* is going down, sometimes I am in hospital.'

In defiance of superstition, his aeroplane had a skull and crossbones, a coffin, two lighted candles and a black heart painted on it. After a British pilot fired on him he so mistrusted the RFC's aircraft identification that he had a red, white and blue V painted on his upper wing between the roundels.

A bad accident in 1915 was followed by four months of hard fighting at Verdun. When he crashed on an air test, he broke both legs and the control column penetrated his palate and dislocated his jaw. When he rejoined his squadron two months later, straight from hospital, he was on crutches but managed to get into his cockpit. A bullet split his lip; he dislocated a knee forced-landing; forced-landing again his machine overturned and broke his jaw for the second time. He did not rise above the rank of lieutenant: although he inspired the other pilots in his squadron, he was not an example to emulate except in courage. In combat, he was like a fighting bull let loose among a bemused *matador* and his henchmen. Astonishingly, he survived the war. It is no surprise that he was France's third-highest scorer, with 45.

An equally ebullient pilot, Jean Navarre, joined *l'Aviation Militaire* on the declaration of war five days after his nineteenth birthday. He began squadron life as a bomber pilot on MF8 and immediately began to exhibit the same brand of recklessness that made Nungesser so conspicuous. The first time he met the enemy, the German flew alongside and waved. Navarre waved back, then put a rifle to his shoulder and fired at him, taking both hands off the joystick and nearly stalling.

Deciding that the Maurice Farman was a platform from which no weapon could be accurately fired, he moved to MS12 and by April 1915 had two victories. Next, he flew for N67 on the Verdun Front. He had red stripes and a skull and crossbones painted on his Nieuport 11, did aerobatics over the French lines and became known as *la Sentinelle de Verdun*. On a dawn sortie he attacked three enemy two-seaters, two of which departed at once. The observer in the third one stood up with raised hands, so Navarre escorted it to his own base as a prize of war. His favourite attack was from astern and slightly below, then he stood up to fire his Lewis gun and could easily have been tipped out. When he had scored seven victories, Nungesser had six. On 4 April he shot down three hostiles on four patrols, but two fell behind German lines and were not counted.

On 17 June he was leading a patrol of three that shot down three Roland two-seaters. He then, from 12,000 ft, spotted another two-seater 3,000 ft below and led the section in a diving attack. To give his companions the best chance of shooting it down, he deliberately turned aside to draw the enemy's fire. One bullet broke a bone in his arm and another wounded him in the side. He fainted but managed to recover enough to make a crash landing. He lost so much blood that he was delirious in hospital for some days and suffered brain damage. This meant that he never flew on operations again. He had a total of twelve victories, which was no small achievement at that stage of the air war.

The records credit him with being the great innovator among French fighter tacticians:

> 'At Verdun Navarre innovated combat between aircraft, and methodical attack. He demonstrated how the little monoplane, obedient to one sole will, could become a dangerous weapon for its adversary, in the hands of a pilot who was skilful, experienced and brave to excess.'

The second-highest scorer, Georges Guynemer with 53 victories, serious, ascetic and religious, was a very different sort from Nungesser and Navarre. In his early youth his parents feared that he was tubercular. When, four months before his eighteenth birthday, France went to war with Germany, he volunteered for any of the armed services that would have him; but looked so frail that he was turned down by all. In November 1914, at his third attempt to enlist, he was accepted as an aviation mechanic. Force of character got him onto a pilot's course and he qualified in April 1915. He began flying with MS3 in one- and two-seater Morane Saulniers. Three months later, with Private Guerder as his observer, they shot down a German reconnaisance machine. Guynemer received the *Médaille Militaire* for this. In November he was credited with bringing down a single-seater. This time, he was made a *Chevalier* of the *Légion d'Honneur*.

By May 1917 he was the longest-serving member of *Les Cigognes*, now flying the Spad 13, and one of the youngest. On 25 May he made four kills. On 11 June, with a score of forty-five, he became an Officer of the Legion of Honour. When August ended he had become the leading French ace, with forty-nine victories. In September he added four more. If his aeroplane was unserviceable, instead of borrowing someone else's, he took any that was awaiting a new arrival in the squadron. These machines were always the worst and often needed servicing before being flown, but he ignored that. One day three of his sorties had to be cut short for this reason and he made three forced landings that the average pilot would not have survived; moreover, in two fights his guns jammed. He flew four sorties of two and a half hours each on four successive days. Photographs of him at that period show him looking gaunt, tired out and ill. On 11 September it was raining but he took off, accompanied by *Sous Lieutenant* Bozon-Verduraz. They attacked a lone two-seater. It was a baited trap: three Albatroses appeared. Bozon-Verduraz fought them and managed to get away safely. He did not see what happened to his No. 1. Days later the Germans informed the squadron that *Leutnant* Kurt Wissemann had shot Guynemer down and killed him.

At the end of the war, the Allied pilot with the greatest number of victories was a *Cigogne*, *Capitaine* René Fonck, with 75. Whereas both Nungesser and Guynemer, like most pilots in all the air forces, were regardless of the amount of ammunition they used in a fight, Fonck, in contrast, told new pilots that he found ten, or at the most fifteen, rounds enough. He also said that if he were ever hit by a bullet he would apply for a transfer to the trenches: in his view, to be wounded indicated a lack of skill. This, of course, was nonsense, as there were so many factors in combat that contributed to victory or defeat. One of the more sensible things he did tell inexperienced pilots was that it was less dangerous to attack fifteen enemy aircraft on their own side of the lines than five behind the German lines. This apparent anomaly was in fact a shrewd observation: he maintained that if one fought several opponents at the same time, they got in one another's way and often held their fire in case they hit one of their comrades. When he engaged a formation he tried to shoot down the leader, then, in the consequent confusion, take out another.

His style was cold, calculating and thorough. To make his aim as accurate as possible, he designed a species of sight, a tube with a plain glass lens at one end, on which were painted two concentric circles. He looked through the other end. The bigger circle covered a field of ten metres at a range of 100 metres. He had memorised the wingspan and fuselage length of every enemy aircraft, so could calculate its distance from the aspect it presented. The smaller circle covered a field of only one metre, which he found useful when very close to his target.

He was criticised for being aloof and conceited, but shyness often gives this impression and it is more likely that there lies the reason for his manner. The French are quarrelsome and critical. Opinions about him were fiercely conflicting in the squadron. The atmosphere became so unpleasant that Colonel Duval, who was the air adviser at General Headquarters, offered Fonck a posting, which he refused. Duval then offered him command of the squadron and transfer of all those who disagreed with his views on air fighting. He asked what would become of the present commanding officer and was told that he would also be transferred. Fonck dismissed this as a mean action, said he would never accept promotion at his own CO's expense and withdrew even further into his shell, sickened by the intrigue. Henceforth he went looking for the enemy on his own.

He was one of only three fighter pilots to shoot down six

aircraft in one day. The other two were Captain J.L. Trollope in March 1917 and Captain H.W. Woollett a month later, both flying Sopwith Camels in 43 Squadron.

Another time, having said he would shoot down five the next day, 9 May 1918, Fonck got up at 4 a.m., took off, found the weather too misty and returned to bed after half an hour. He was roused again at noon, found the mist still thick so had lunch and waited until the sun appeared at 3.30 p.m. Accompanied by two others, he took off a quarter of an hour later, when the mist had cleared. After thirty minutes a Rumpler escorted by two Halberstadts came in sight. He told his wing men to keep back, shot down one Halberstadt and then the Rumpler. It was now 4.20 p.m. Presently he returned to base to refuel and rearm, then took off again accompanied by two different pilots. He shot down an Albatros at 5.17 p.m. Next, several aeroplanes appeared in the distance. Closing them, he saw that they were five Fokker DVIIIs in V formation and four Pfalz DIIIs in diamond at a lower level. He stalked them through cloud, shot down the rearmost Pfalz, then, from forty metres' range, shot the leading Fokker down with only four rounds of ammunition.

When Fonck wrote his autobiography after the war, he maintained that to be successful a fighter pilot must know how to control his nerves, how best to attain absolute mastery of a situation and how to think coolly in difficult ones. He also shared a guiding principle with Manfred von Richthofen, who asserted that a fighter pilot must go into action telling himself that his opponent '*must* fall'. Fonck put it: 'I always believed that it is indispensable to maintain absolute confidence in ultimate success, along with the most complete disdain for danger.'

He had more detractors among his comrades than any other famous Allied or German pilot. His comrades sneered at him for opening an attack only when he was in a position of such advantage that he was certain of victory and of no injury to himself. The truth is that he was a decent man who showed his loyalty, high personal standards and repugnance for intrigue when he refused the offer to supplant his commanding officer. As for the accusation of fighting only when in a position of supreme advantage, the words of a Second World War fighter leader puts that in its true perspective.

Squadron Leader Peter Blomfield DSO, DFC, who commanded 260 Squadron in Desert Air Force, said about Squadron Leader Osgood Hanbury who was commanding the squadron when 'Blom' joined it:

'He gave the impression of being as mad as the proverbial hatter on the ground, however a cold, efficient leader in the air. I flew as his Number Two for the next two months. What I learned in that short time probably enabled me to stay alive in spite of all the Me109s could do and to remember his oft-repeated adage, "He who fights and runs away lives to fight another day; the fool who stays and takes a chance gets took home in the ambulance". No wonder we all loved him.'

When one seasoned fighter pilot says that about another, he confirms the good sense of Fonck's principles.

These sentiments were an endorsement of Manfred's declaration that a pilot who chased an adversary deep into enemy territory or in any other way took unnecessary risks was likely to be killed early in his career. He was less useful to his country than a prudent pilot who, although he fought boldly, knew when to abandon a fight and would live to shoot down many more of the enemy.

Maggiore Francesco Baracca, with thirty-six victories the leading ace of Italy's *Aeronautica Militare*, was twenty-three when Italy entered the war on the Allied side. He was a regular officer who had transferred from the cavalry in 1912. With several other Italian officers he did his pilot training at the Reims flying school, where his instructor reported, 'He has sensibility, sharp sight, control over the nerves. He is undoubtedly a first-class pupil, the best of the Italian party.' Two days before Italy declared war, he and five other pilots were sent to France for instruction on the Nieuport-Macchi (Nieuport 10 manufactured under licence by Macchi). On 19 July 1915 he joined No 1 Fighter Squadron. During the next four months he was in action three times;

unsuccessfully, because his gun jammed during each fight. This was not its only defect, as his diary shows:

'My machine-gun is a new weapon. We do not know it well and the fault is to some extent ours. It is badly placed in the aeroplane and to shoot is a very acrobatic business, and I lost faith in being able to do anything.'

By the end of March 1916 the Italian air force strength was seven Caproni bomber squadrons, two Voisin and eight Farman reconnaissance squadrons, five Caudron and two Farman artillery observation squadrons, one seaplane squadron and five fighter squadrons. This was meagre in comparison with Britain, France and Germany. The official history informs us, with Latin embellishment,

'The fighters were born in the spring of 1916. They were given their baptism of glory by Captain Francesco Baracca, who initiated his series of victories by bringing down two aeroplanes adorned with the enemy's insignia of the black cross, in the sky above Medeuzza.'

Italian aircraft bore red, white and green roundels; the Austrian, a black cross and broad red-white-red stripes on the wingtips and fuselage.

At first light, 4 a.m., No. 1 Squadron's fighters took off to intercept enemy bombers that were under anti-aircraft fire and searchlight illumination, climbed to 2,000 metres and dispersed to operate singly. In a letter to his family, Baracca described events.

'I saw above me the big wings of an Aviatik [BIII, which did not handle well and was known as "the rocking chair", but had a machine-gun for the observer]. He was going fast [about 70 mph] and I gradually gained on him. When he climbed, I increased speed. Drawing close, I had begun a most difficult manoeuvre to protect me from his shots. I saw the machine-gunner aiming in one direction and I veered in another, then vice versa. This game continued for several minutes until I was positioned fifty metres behind his tail at a height of about 3,000 metres. Then, in an instant, I aimed and fired 45 rounds. A moment later the enemy swerved heavily to one side and was thrown almost vertical.'

The twenty-four-year-old pilot of the Aviatik was wounded in the head, the petrol tank was riddled, the observer had slumped over his gun, screaming and shedding blood, but the pilot made a controlled landing in a field. Baracca landed near him and was hoisted shoulder-high by Italian soldiers.

Baracca scored again on 16 May, 23 August and 16 September. He did not shoot down a fifth and qualify as an ace until 11 February 1917, which shows how small the scale of air fighting on the Italian/Austro-Hungarian Front was compared with France.

His method of attack when he met an enemy formation was to select the aircraft that was worst placed for protection by the others' guns. He would then make climbing turns until high enough to launch an attack and hold his fire until he had closed the range to between twenty and fifty metres.

Early in June, with thirteen victories to his credit, he took command of No. 91, 'the aces' squadron', flying the Spad. Other future high scorers serving under him were *Tenente* Flavio Baracchini, who shot down twenty-one, another lieutenant, Prince Rufo di Calabria, whose final total was twenty, and Lieutenant Ranza, who achieved seventeen.

The second-highest Italian scorer was *Tenente* Silvio Scaroni, with twenty-six, *Maggiore* Pier Ruggiero Piccio was next with twenty-four.

On 19 June 1918, Baracca was killed while strafing trenches with two other pilots. Piccio was shot down and taken prisoner on 27 October 1918.

A report by the Austrian Second Army obtained by the Italians at the end of the war, says.

'The superiority of the enemy's aeroplanes is indisputable, both in numbers and quality. The opinion of our troops and our High Command is that the enemy flyers were bold, reckless and resolute, with a rare offensive spirit.'

In addition to the five most successful fighter pilots, there were two who scored seventeen, one with twelve and two with eleven victories. Six had eight victories, five had seven, ten had six, seven scored five each.

The total number of Austro-Hungarian aircraft shot down in 1915 was six; in 1916, forty-eight. During November and December 1917, four RFC squadrons arrived in Italy: Nos 28, 66 and 45, all flying Camels, and 34 with RE8s. Between 29 November and 31 December they shot down fourteen hostiles out of a total 231. The following year the RFC and Italian air force together brought down 647. Austria-Hungary's victories were two in 1915, fifteen in 1916, fifty-five in 1917 and thirty-two in 1918.

The outstanding RFC pilot fighting in Italy was Captain (later Major) William Barker VC, DSO, MC, a Canadian in 28 Squadron. He had been a high scorer in France, returned there in June 1918 and ended the war with fifty-three victories.

The Italians did not contribute anything significant to the advancement of fighter tactics, but the enemy's estimation of them was well merited.

Although the United States of America did not enter the Great War until 6 April 1917, American fighter pilots had been serving in *L'Aviation Militaire* since 1916. Soon after the war began, Norman Prince, a private pilot, had arrived in France to form a volunteer American squadron. With the help of Dr Edmund Gros, who had formed the American Ambulance Service, and the millionaire William Vanderbilt, Americans who had already joined the French Army were recruited. The unit was not formed until 16 April 1916, under a French commanding officer, *Capitaine* Georges Thénault, and second-in-command, Lieutenant de Laage de Meux, with seven American pilots, one of whom, Bert Hall, was already in *L'Aviation Militaire*. They were provided with Nieuport 11s and given the number N124. The squadron was originally called *L'Escadrille Americaine*, but the Germans objected because the USA was neutral, so the name was changed to *Lafayette*. It fought with distinction but was disbanded on 18 February

1918 and the American pilots were transferred to the United States Air Service.

Some Americans were serving in the RFC or RNAS. The most successful of these were W.C. Lambert, who shot down twenty-four enemy aircraft, S.W. Rosevear with a score of twenty-three and three pilots who each had twenty: J.J. Malone, F.W. Gillette and G.A. Vaughan. None of these transferred to the United States Air Service.

One of the former *Lafayette* members, Raoul Lufbery, was appointed commander of the 94th Pursuit Squadron. He became the USAS's third-highest scorer, with seventeen victories. The second highest, who had not flown with the French, was Frank Luke, who scored twenty-one. The top American ace was Edward Rickenbacker, who began learning to fly in January 1918. He shot down twenty-six aeroplanes and balloons between 29 April and the armistice on 11 November that year. It was a remarkable achievement to take out his first enemy aircraft after only four months' flying experience, when they were much harder to hit than ever. To amass a total twenty-six in a trifle over six months, with less than a year's flying experience, needed phenomenal natural talent. Rickenbacker had been a champion motor racing driver, which meant that he had excellent reflexes, an essential factor in his temporary occupation of fighter pilot.

Chapter 13

A Rest, then Manfred Returns to the Front

The honours lavished on Manfred during his leave far surpassed those accorded to the heroes of any other nation. For the British and members of the Empire armed forces, the summit of recognition was an invitation to an investiture at Buckingham Palace, when King George V pinned medals on tunics and uttered brief congratulations, without a handshake. In France and Italy such ceremonies were marked by laudatory speeches and kisses on both cheeks. In Germany the bestowal of decorations was normally as simple as in Britain.

When Manfred went on leave on May Day 1917, it was at his monarch's orders. Told not to fly himself to Bad Kreuznach, where Supreme Headquarters was situated, he flew as passenger in a two-seater piloted by *Leutnant* Krefft, who was going on sick leave. Even of this, the Kaiser and his Staff would not have approved: he had been told to travel by train, but time did not allow.

They landed at Cologne for lunch, where a crowd waited to welcome Manfred, and arrived at their destination that evening to another enthusiastic reception. Manfred was taken to meet General von Hoeppner, Commander of the *Luftstreitkräfte*. He also met the Prince of Pless, who invited him to hunt bison. Next day, his twenty-fifth birthday, he lunched with the *Kaiser*, who congratulated him on his fifty-two victories and gave him a present, a bronze and marble bust of His Majesty. He also urged him to take care of himself and asked his ADC why

Manfred was still flying. The answer was that he was needed not only as an example to others but also for his own prolific contribution to the destruction of enemy aircraft and the men in them. After lunch the *Kaiser* spent half an hour in conversation with his guest of honour. Next day Krefft and Manfred flew on to Bad Homburg, where the *Kaiserin* greeted them at the airfield and presented Manfred with a gold cigarette case. In the evening he was guest of honour at a banquet at which *Feldmarschall* von Hindenburg presided.

On 23 May he had written to his mother, 'I intend to come home at the beginning of May, but before that I will go pheasant shooting, to which I have been invited and to which I am looking forward eagerly'. On 4 May, he began a shoot in the Black Forest that lasted several days. On 9 May, in a letter to her he confessed,

'I suppose you will be cross with me for having been in Germany for eight days without writing to you. I am shooting pheasant here and expect to stay until the fourteenth. The sport is excellent. After that I have to go to Berlin for three days to examine some new aeroplanes, after which I shall come to Schweidnitz. From there I'll go on to the Prince of Pless's estate to shoot a bison. Then I have to visit other war fronts, which will take three or four weeks'.

The so-called bison was a species of wild ox found only on the Pless estate and in the *Czar* of Russia's Bielowicz Forest. His account of his big game hunt said,

'I arrived at Pless in the afternoon of 26th May, impatient to kill a bison that evening. An hour's drive and thirty minutes' walking took us to the scene of the shoot, where the beaters were ready and now began the drive. Suddenly I saw a monster coming towards me through the trees. I had hunting fever. It was a mighty bull, two hundred and fifty paces away. I was too far to shoot. I might have hit it, for it was so big, but searching for it would have been unpleasant and to miss would be a disgrace. I waited until it came closer.

He probably scented the beaters, for he turned towards me at a speed remarkable for so big a beast, but at a bad angle for a shot. He disappeared behind some trees. When he charged towards me I had the same feeling that grips me when I see an English aeroplane. A second bison appeared and I shot it at a hundred paces. I had to hit it twice more before wounding it mortally.'

On 4 May Manfred had a telegram telling him that Lothar had been awarded the *Pour le Mérite*. On 13 May he received another, to inform him that Lothar had been wounded, but not mortally so. He had scored his twenty-fourth victory, another easy one against a BE2. The wound was on his hip and kept him in hospital and on sick leave for five months.

While Manfred was in Berlin a publisher offered him a contract for his memoirs; he was photographed, as all holders of the Blue Max were, for a postcard that would sell in thousands; and he was involved in discussions about rectifying the weakness in the Albatros DIII's lower mainplane. There was also a new fighter, the Roland DIII, to test. When he finally reached home on 19 May word spread fast and bouquets and presents began arriving at the house. Such attentions were not to his taste but he had to put up with several days of adulatory welcome and speeches in his honour.

His publisher sent a shorthand typist to work with him on his book, which occasioned a display of his sense of humour. They were in the garden, near the front gate, when two women paused on the road to speak to him and satisfy their curiosity about his pretty companion. He mischievously introduced her as his fiancée, to her embarrassment.

On the last day of May two aeroplanes landed on the parade ground outside the town. One was a Halberstadt scout in which Manfred was to set off on his trip to Austria, the Balkans and Turkey, and a two-seater in which the pilot who had flown the Halberstadt would depart as passenger. He told Manfred that there was no need to fasten his seat belt, as the aircraft was so stable. Either he was an idiot or meant this as a jest, exaggerating the single-seater's handling qualities. It was common practice to trim the Albatros and other types to fly hands-off, but Manfred replied that he always buckled his safety belt. On the way to his first refuelling stop, he did release the joystick, the Halberstadt half-rolled and he would have fallen to his death had he not been strapped into his seat.

On the British front, the Germans held a commanding position on Messines Ridge, a fortified salient from which they could see the trenches and forward batteries. At 3.10 a.m on 7 June, after weeks of tunnelling, nineteen mines were exploded simultaneously under the ridge and an artillery bombardment began. Soon after, the infantry advanced and within three hours the ridge was in British and New Zealand hands.

During the first week of June the RFC and RNAS fighters had demonstrated such superiority over the *Luftstreitkräfte* that Manfred was urgently needed at the front. The Germans' defeat in the Battle of Messines made his return to *Jasta* 11 all the more imperative. His goodwill tour was cancelled and he was summoned to Supreme HQ, and a banquet with the *Kaiser* in honour of the *Czar* of Bulgaria. He broke his journey there in Berlin, to emphasise to the Military Aviation Inspectorate the need to cure the Albatros DV's lower mainplane weakness and to produce a new and better fighter to replace it. After his stay at Headquarters he visited Lothar, who was convalescing in Hamburg, before resuming command of his squadron, which had moved to Harlebeke, near Courtrai.

On 18 June he recorded his fifty-third victory by destroying an RE8. Five days later he claimed a Spad: although it was allowed, the best authorities are agreed that Allied records do not substantiate this. On 24 June he shot down a reconnaissance DH4, one of a pair escorted by ten scouts.

At Supreme HQ he had been told that the first fighter *Geschwader* was to be formed. It would consist of four *Staffeln*, corresponding more or less to a French *Groupe* and a British Wing. He was to be given command of it, with authority to appoint the *Jasta* commanders. On 24 June he was confirmed as *Geschwader Kommandeur* of JG 1, comprising *Jasta* 4, 6, 10 and 11, each with an establishment of twelve aircraft and twelve pilots. The JGs were intended as mobile tactical units sufficiently strong to obtain dominance in the air over any sector of the front line. *JG1* was allotted to the Fourth Army.

Ideally, all the *Jastas* would share the same airfield. Manfred's assembled around Courtrai and each had a distinguishing livery: No. 11, at Marckebeke, already had red; No. 10's, at Marcke, was yellow; No. 6's, at Bisseghem, black and yellow stripes; No. 4's, at Marckebeke, a wavy black stripe along each side. An estate owned by Baron Jean de Bethune at Marcke, from which the neighbourhood took its name, Marckebeke, had been requisitioned as *JG* 1's Headquarters. In the castle there, Manfred installed the staff officers of both *JG* 1 and *Jasta* 11, which he still commanded. In this way a *JG* differed from a *Groupe* or Wing, whose commanders were unfettered by squadron responsibilities. Understandably, the Baron resented having under his roof any specimens of the *boches* who were ravaging his country. He had the courage to show his animosity by his demeanour and by denying them the use of several rooms. The latter was short-lived; the *Freiherr* ordered him to unlock those that his staff officers needed.

Manfred now had the first chance to prove his organisational and administrative ability and operational breadth of view. He acquitted himself efficiently when he briefed his squadron commanders on the arrangements he had made. Communications came top of the list: instead of having to wait for reports on enemy aircraft movements to be passed via a central point that supplied this information, sent by spotters at numerous vantage points to several airfields, the reports would be given directly to *JG* 1's HQ; and a telephone link now enabled all the squadrons to share simultaneously in calls from him and his staff. He put his squadrons on a roster by which they took it in turns to be at readiness for take-off at set times from dawn to dusk.

To use World War Two military jargon, he also put them in the picture about the RFC's general air activity over their sector of the line. Ground strafing had been carried on since 1915, when fighters returning from a sortie with all or some of their ammunition left often flew low along the trenches firing their guns. By 1916 the RFC was doing this so often that a German officer wrote about those times.

'The infantry had no training in defence against very low-flying aeroplanes. Moreover they had no confidence in their ability to shoot these machines down if they were determined to press home their attacks. As a result they were seized with a fear amounting almost to panic; a fear that was fostered by the incessant activity and hostility of enemy aeroplanes.'

The Germans were also doing a lot of ground strafing by mid-1917, with purpose-designed aeroplanes formed into *Schlachtstaffeln, Schlastas* for short. They were all two-seaters and began with the AEG CIV, which had an armoured belly and was armed with a Spandau and a Parabellum. The Hannover CLIII, weighing 2378 lb, was also armour-plated underneath and had a Spandau and a Parabellum. The Junkers JI, with two Spandaus for the pilot and a Parabellum for the observer, had so much armour that it weighed 4787 lb. The Halberstadt CLII carried a Spandau and a Parabellum, had an armoured floor and, weighing 2532 lb, was the most manoeuvrable in avoiding ground fire. Neither the British nor the French had special ground-attack aeroplanes.

Manfred also painted a grim picture of big bomber formations swarming over German territory, which his fighters must intercept.

His own *Jasta* was at readiness on the morning of 6 July and ordered up to deal with ground strafers over the trenches. They were at 12,000 ft, which suggests that their original targets had finished their work and gone home before *Jasta* 11 could tackle them, when some FE2es of 20 Squadron came in sight. He heard Kurt Wolff, his No. 2, open fire just before he himself was attacked. The Fee's observer started shooting at such long range that Manfred had not yet cocked his guns, because he did not think he could be hit from 300 yards. The Fee turned head-on and Manfred intended to turn in astern as soon as it next changed direction.

'Suddenly I felt a blow on my head. I had been hit. For a moment I was completely paralysed. My hands dropped to my sides, my legs were limp. The worst part was that the blow on my head had affected the optic nerve and I was completely blinded. The machine dived and I doubted that the wings could stand it.'

He was still conscious but unable to take hold of the control column. He managed to switch off the fuel and ignition. He could see nothing, not even the strong sunlight penetrated his eyes. When he was down to an estimated one or two thousand metres, he found his vision returning. He had some control over the Albatros but it kept steepening its glide into a dive. When he could read the altimeter it showed 800 metres. He restarted the engine and looked around. There were shell holes in every direction, but he saw a forest that he knew was behind German lines. His pilots had followed and he was soon down to fifty metres, but still there were only shell craters in which to land. He decided to fly eastwards for as long as he remained conscious. Presently he started blacking out again, so crash-landed near a road, tangling with broken telephone wires. Unable to climb out, he had to wait until help arrived. His pilots also put down near him. Soldiers appeared, recognised his aeroplane and he was soon in hospital at Courtrai. 'I had quite a respectable hole in my head'. The wound was about two inches long and exposed his skull, which was fractured.

He admitted that his immediate anxiety was lest Lothar were passed fit for flying before him.

Nobody claimed to have shot the Red Baron down, but the evidence points to second Lieutenant A.E. Woodbridge, an observer who fired several bursts at him. The pilot, Capt C.A. Cunnell, was killed in action six days later when flying with a different observer who flew the aircraft back to base.

Interviewed about the fight eleven years later, Woodbridge said:

'Of course, neither Cunnell nor I had any idea that it was the Baron himself with whom we were exchanging pot shots, and, for that matter, I don't know it yet, but I must say that Richthofen's account of the fight coincides exactly with our reports made at the time and my recollections of that damned busy forty minutes we had with his flock of red Albatros scouts.'

This appears to bear out that all *Jasta* 11's aeroplanes were now painted solid red, in which event the original purpose, to make their leader easily recognisable, must have been abandoned.

CHAPTER 14

THE FOKKER TRIPLANE
ENTERS SERVICE

If Woodbridge's bullet, fired from a range at which Manfred thought he was safe, had given him his final comeuppance, the lives of sixteen Britons, the wounding of four and capture of eight might not have happened; but, given the frequency and ferocity of the air battles and the skill of many other German pilots, many of these inflictions were probably inevitable anyway.

The severe head wound had a profound effect on Manfred's brain. His head had to be shaved and a bandage wrapped round it that reached down to his ears. The alteration to his appearance was temporary, whereas the change in his behaviour was permanent. As the effect of the anaesthetic faded, his head began to ache and for the few months that were left to him the pain would constantly recur and affect his conduct.

Like all good commanders in every war, on land, at sea or in the air, he had the devotion of his officers and men. More than that, he had become an icon held in awe and affection as well as admiration. His Administrative Officer, *Oberleutnant* Karl Bodenschatz, and three of the squadron commanders visited him in hospital as soon as they were allowed and had to answer a barrage of questions about his welfare when they got back. *Oberleutnant* Kurt von Döring was temporarily commanding JG1 and *Leutnant* Kurt Wolff had taken over *Jasta* 11.

When Manfred arrived home for a few days his mother was shocked to find that his wound had not yet healed and he had to go to a clinic every day for the dressing to be changed. 'He does not look good and is irritable' she wrote in *Mein Kriegstagebuch* (My War Diary). She asked him not to fly any more, to which he replied, 'Who would fight the war if we all thought like that? The soldier in the trenches? When the professional fails in leadership, the situation soon becomes as it is in Russia.'

She reminded him that in the Army everyone was given regular rests behind the lines, whereas he fought every day. His retort was, 'Would you be happy if I were to rest on my laurels?'

One evening when his persistent headache drove him early to bed, a group of locals called to pay him homage. His father felt duty-bound to rouse him. He reluctantly got up to meet them in a vile mood plain to everyone. After the intruders had left, Kunigunde *Freifrau* von Richthofen asked him to try to be friendly on such occasions in future. He told her that his best reward was not demonstration of admiration by civilians, it was when infantrymen, regardless of danger, climbed out of the trenches to 'shout joyfully at me and I look into their grey faces, worn by hunger, battle and loss of sleep'. There is an honest depth of compassion there that he would never have expressed to anyone but his immediate family.

Another secret he confided to his mother was that he had for a long time corresponded with a girl whom he wished to marry, but 'not as long as I am liable to die any day'.

Manfred's absence did not immediately affect JG1's fortunes. On the first day under his deputy's command, it shot down ten RFC machines in the course of three patrols: on each of which the Albatroses considerably outnumbered the Sopwith Triplanes and 1½-strutters they fought. During the next few days Wolff was wounded and had to hand over command to a successor who had not yet won any victories but was his senior. Air activity was increasing and JG1's losses mounting. Von Döring kept Manfred in touch with events. His tactic of sending up big formations had been dropped by a more senior officer at 4th Army HQ in favour of patrols in squadron strength. Moreover, these were confined to a specific area, like the old-fashioned barrier patrols. Fighter tactics had changed considerably and this was a retrograde step ordered by a Staff officer who knew as much about the subject as a Tibetan lama would know about ballroom dancing. Outraged,

Manfred wrote in protest and, at the same time, complained about the quality of the current German fighters. The Sopwith Triplane, Spad 7 and Sopwith Camel were not only superior to the Albatros DV but also outnumbered it.

On 25 July Manfred returned to JG1, although still grounded, bringing the good news that the *Geschwader* was about to receive Fokker triplanes, the latest scouts, with, he was promised, a phenomenal rate of climb and superb handling qualities. Despite his stricture of the Albatros, there was information to gladden him: in the last few days JG1 had shot down a Sopwith Triplane and a Camel.

It was his expressed principle that, in the air, the commander of a *Geschwader* should not boss his squadron commanders around too much. They must have unconditional freedom in the area in which he has ordered them to operate. In other respects, he was stern as well as sensible. Dissatisfied with *Jasta* 10's commander, who ignored instructions always to fly in numbers and used to go hunting alone, he replaced him with twenty-year-old Werner Voss, who had rapidly built up a score of thirty-four kills. Voss had begun by enlisting in the hussars, transferred to the air service as an observer, qualified as a pilot and served with Manfred in Boelcke's *Jasta* 2. His pleasant personality endeared him to his comrades and he was as merciful as possible to the enemy. When he left his bomber squadron he was the only survivor of all the pilots and observers who had been on its strength when he joined. Seeing so many of his friends shot down in flames, he had such sympathy for all two-seater crews that he habitually fired at the engine in order to give them a chance of survival. This was genuine chivalry. He was a Jew – lucky that he was destined to die before Hitler got his hands on Germany.

The RFC let Manfred know that they were aware of his return to the fray: No. 100 Squadron bombed the Marcke airfield. He was still grounded.

The Allies' capture of Messines Ridge was the preliminary to the Third Battle of Ypres, launched on 31 July 1917 with the customary artillery barrage. British air activity concurrently became more intense, but Manfred, not wishing to commit his

Albatroses in large numbers to do battle with Sopwith Triplanes, Spad 7s and Camels, temporarily sent up one squadron or fewer at a time.

On 16 August, leading a formation of five in his favourite formation, an arrowhead with two aircraft echeloned on each side and stepped up from front to rear, he flew for the first time since being wounded. They intercepted some Nieuport 7s; he sent one into a spin by shooting it in the engine and fuel tank, followed it down and, with unnecessary savagery, gave it a final burst that sent the already doomed machine plunging into the ground. The sortie wearied him so much that he went to bed.

Four victories on the next day, in which he did not share, brought *Jasta* 11's total to 200 and Manfred ordered a bottle of champagne: a rare event, as, in his view, alcohol affected fitness to fly, so there was little drinking in his *Geschwader*. The double century elicited a congratulatory telegram from General Hoeppner. This encouraged Manfred to write to HQ 4th Army, under whose orders *JG*1 operated, and point out that his pilots were overburdened. After having made long flights escorting bombers or ground-attack aircraft, they were tired and should not be required for combat patrols on the same day. This was heeded and Hoeppner signalled him again, expressing the wish that he should not fly unnecessarily until completely recovered from his wound.

On the morning of 26 August Manfred took advantage of a justifiable reason for flying. At the head of a five-formation, he caught up with a lone Spad 7 at 3,000 metres, which he thought must be seeking a low-flying German artillery spotter.

> 'When he came out of sun I attacked him. He tried to escape by diving, I shot at him and he disappeared in cloud. I followed him and saw him explode at about 500 metres altitude. The new very bad incendiary ammunition did a lot of damage to my pressure and intake pipes, etc and it would have been impossible to follow a slightly wounded adversary.'

What a pity that it didn't do fatal damage and kill him: anyone defying the Geneva Convention and using the bullets that

Mannock had bitterly condemned, deserved to die in flames himself.

He had still not recovered physically or mentally from his wound, as revealed in a letter to his mother two days later.

'I am glad to hear that Lothar is continuing to improve, but he should not be allowed to return to the front before he is fully fit. If he is allowed to do otherwise, he will suffer a relapse or be shot down. I speak from experience. I have made only two combat flights since my return. Both were successful, but after both I was completely exhausted. During the first one I was nearly airsick. My wound is healing very slowly and still as large as a five-mark piece. Yesterday they removed another splinter of bone, which I think will be the last.'

On 28 August two Fokker Triplanes were delivered and Voss took one up to demonstrate it to the other pilots. On 1 September, he and Manfred each flying a Triplane, accompanied by three Albatroses, caught an RE8 whose pilot and observer probably assumed the triplanes to be RNAS Sopwiths. Manfred shot it down, his 60th victory, which would be celebrated by the last of his commemorative cups: the jeweller in Berlin informed him that no more silver would be available; another wartime dearth.

On 3 September, flying his triplane at the head of four Albatroses in V formation, he fought a flight of Sopwith Pups. This time, he did not display the merciless determination to kill that lay deep in his character, but allowed his adversary to crash land from respect for him. The Pup pilot kept shooting right down to fifty metres altitude, strafed a column of infantry in the last few seconds before flaring out to make his landing, then deliberately smashed his aeroplane by ramming a tree.

Manfred now went on four weeks' leave, during which, of course, he spent most of his time shooting birds and four-legged game.

The Richthofen brothers beside a Fokker Triplane. Lothar is on the left and Manfred on the right, seen carrying the *Geschwaderstock* under his left arm. It was virtually the badge of office of the *Jagdgeschwader* 1 Commanders.

Major Albrecht *Freiherr* von Richthofen with Lothar and Manfred.

Manfred with a bandaged head
wound meets the Kaiser.

Manfred and *Leutnant* Krefft
with ladies of the Empress's
retinue.

Hauptmann Hans-Joachim Marseille.

Oberst Werner Mölders.

Hauptmann Max Immelmann.

Hauptmann Oswald Boelcke.

Major James McCudden VC DSO MC who began the war as a RFC mechanic.

Capitaine Georges Guynemer.

Seen standing on the left is Lieutenant Raymond Collishaw, the RNAS's top-scoring fighter pilot.

(*Above left*) *Leutnant* Werner Voss (left) and Anthony Fokker the Dutch aircraft designer forced to work in Germany (centre).

(*Above right*) *Capitaine* Jean Navarre.

(*Left*) Captain J. L. Trollope.

Capitaine René Fonck, Allied fighter pilot.

Fokker Triplane flown by
Manfred von Richthofen.

(Above) Manfred von Richthofen taking off.

(Below) Albatros DVs on a front-line airfield.

Leutnant Werner Voss (right) and his Fokker Triplane.

Manfred von Richthofen's Albatros DIII.

Sopwith Camels of No 8 Squadron RNAS, later No 206 Squadron RAF.

SE5A with BE2es in the background.

America's leading ace Major Eddie Rickenbacker and his Spad 13.

Vickers BE2c, the RFC's latest type at outbreak of war.

Chapter 15

In Action Again and the Fokker Triplane's Defects

While Manfred was on leave, Wolff, with 33 victories and Voss with 48 were killed.

On 25 September Lothar rejoined *Jasta* 11 and took over from its acting commander, *Leutnant* Groos, who had been wounded eleven days earlier but refused to go to hospital. Thenceforth Manfred, wondering how fast his brother was catching him up in the victory stakes before he could also return to the fray, must have been as uncomfortable as a nudist sitting on an ants' nest.

His game-hunting holiday ended on 30 September. He wrote to his mother to say he was very glad to hear of Lothar's sudden recovery and looked forward to flying with him again and giving the 'English' a hard time together. He was no less pleased about his 'sporting' activities:

> 'My bag during the past fortnight has not been bad, a large elk, three excellent stags and a buck. I am quite proud of my record, as Dad has shot only three stags in his life. I am going to Berlin today and will be with you within a week.'

Although he was proud of the place he held in the esteem and affection of his compatriots and appreciative of the public demonstrations that welcomed him wherever he went, he never became swollen-headed or showed off. Despite his enormous innate self-confidence, he was essentially shy and guarded about his personal life. Although he had enjoyed sharing in the convivial side of life in an officers' mess, when he became an airman he immediately recognised that carousing was incompatible with flying.

Going about his leisure activities, he tried to be self-effacing. On the way by train to Berlin and the next appointment of his leave, he shared a compartment with a civilian who described how Manfred turned up his greatcoat collar and went to sleep. When he woke and was recognised he chatted unaffectedly for the rest of the journey and, before alighting at their destination, hid his Blue Max under his tunic so that he would not attract attention.

A military friend who often entertained him at home said that although his manners in mixed company were impeccable, 'he was not a ladies' man in the usual sense of the term', although women clearly found him attractive. Mobbed by them at a race meeting, he politely signed the race cards they thrust at him, whereas most other celebrities would have been annoyed. 'He never put on airs or posed.'

A Fokker Triplane direct from the factory awaited his arrival back in command of *JG*1 on 23 October. His pleasure at this was dampened by the events of the next few days. First, there was an incident of the 'friendly fire' that has always been a hazard of war and has become an increasingly frequent occurrence in the present century. One of his pilots was shot down and killed by an unidentified 'friend' who obviously mistook the Fokker Triplane for the more familiar Sopwith 'Tripe'. Next, when the Richthofen brothers took off the following morning for their first patrol together for nearly half a year, Manfred noticed that Lothar's triplane was flying in a highly irregular fashion that compelled him to cut the engine and glide back to the ground. Manfred followed him and was astonished to see that, although the machine landed safely, it immediately started to break up; a dire warning about its structural defects. Next day, another *Jasta* 11 pilot was killed in a crash caused by the collapse of his top wing. On 2 November all Fokker Triplanes were grounded and an investigation into the causes of the accidents began. The German aerodynamicists were plainly inferior to the British and

French. Moreover, this type was not an original Anthony Fokker design, but merely an attempt to copy the Sopwith.

For the time being, the *Geschwader* flew only the Albatros DV and the recently received Pflaz DIII, neither of which was a match for the Camel, the Sopwith Triplane or, in the hands of above average pilots such as 56 Squadron's, the SE5A. The wings (which were still likely to collapse) of Manfred's Albatros DV, tailplane and fin were now the only components bearing his distinctive red paint.

Fog, rain and low cloud hindered flying for several days, during which the whole *Jadgeschwader* moved to airfields near Cambrai, where the battle of that name had begun on 20 November with a surprise attack by British tanks and infantry, not heralded this time by a long artillery bombardment.

During the next seven days Manfred sent down his sixty-second and sixty-third British aircraft. One of his pilots, *Leutnant* Hans Klein, became the twenty-ninth fighter pilot to win the Blue Max, the seventh under his command to be thus honoured.

Winter weather frequently grounded the aircraft on both sides of the lines. Although Manfred detested inactivity, it was obviously balm for his troubled body and mind. Describing his feelings after a fight, he said, 'I am in very low spirits. But that is no doubt an after effect of my head wound'. When he landed he went to his room and did not want to be disturbed by anybody for any reason. He said he pondered on the war as it really was; not, as the public imagined, a matter of triumphant cheering and jubilation. 'It is very serious and very grim.'

In early December he went to Berlin for ten days, during which he tested the prototype Pfalz Triplane. It was a disappointment, its performance worse than the Fokker's.

In a letter to his mother dated 11 December 1917, he said that little was happening at the front, so life was rather dull. He intended to spend Christmas with his squadron, which of course included Lothar, and their father. *Major* von Richthofen was a familiar and popular figure in the mess, where he greatly enjoyed listening to the flying talk which is and always has been the main topic among pilots and air crew. Manfred also mentioned that his batman had posted a parcel for Bolko and he hoped its contents would appeal to his tastes.

In Russia, the Communist revolution in December resulted in an armistice on the Eastern Front. A peace conference was to be held at Brest-Litovsk after Christmas. Invited by Prince Leopold of Bavaria, who commanded the German forces in the east, Manfred and Lothar set off for Brest-Litovsk to attend the peace conference and to shoot big game. Manfred deplored the fact that the herds of 'bison' that had numbered a good 700 head before the war had diminished to an estimated 150. Hungry soldiers were to blame. Consequently, not wishing to reduce the survivors by more than a couple of head, the brothers contented themselves with one stag apiece.

On 15 January 1918 Manfred wrote to their mother to explain why she had not heard from him for some time. 'The stay in the quiet forest has done us both a great deal of good. I am often in Berlin and shall be there for a fortnight from the 20th and hope I shall see you.'

There had been strikes in the munition factories, so during his stay in the capital one of his duties was to visit them and tell the workers how important their contribution was to the war effort. He and other experienced fighter pilots also flight tested several prototype aircraft and decided which to abandon and which to develop. He was an innovator in this by ensuring that the trials were not left to factory pilots or others who had not seen action.

At the end of the month he flew from Berlin to Schweidnitz, overflying Wahlstatt, where his youngest brother was a cadet. The boys happened to be on parade, so he succumbed to temptation and performed a loop over the square. When Manfred left Schweidnitz, Lothar, who had been suffering from ear trouble, stayed behind and Manfred urged their mother to ensure that he did not return to his squadron until the end of the month. On the flight back he 'buzzed' the cadet school, which brought its inmates out to collect the boxes of chocolates he threw out.

The latest batch of Fokker Triplanes appeared safer than the earlier ones, as the wings had not been breaking; until, on

2 February, part of a top wing did collapse and the pilot concerned was lucky to get away with his life in a forced landing. Another worry was the engine; a rotary, it did not give enough power. Manfred complained that not only was an in-line engine, preferably a supercharged one, more efficient, but also the synthetic oil used in rotaries was of poor quality.

His almost four months of varied activities, travel and no fighting ended on 12 March 1918 when, accompanied by Lothar and *Leutnant* Steinhäuser, he met eight Bristol Fighters that were patrolling between Cambrai and Cautry. Lothar quickly shot down two, one in flames; Steinhäuser got one; and Manfred forced down a fourth, whose crew were taken prisoner, after fifteen minutes' fighting in which he shot off half the observer's left arm.

Next day, Manfred led three *Staffeln* in one of which Lothar was flying, in an attack on twelve DH4 bombers that had an escort of eleven Bristol Fighters and twelve Camels. Early in the battle Lothar's top mainplane was destroyed and his rudder damaged. Unable to maintain altitude or steer, he withdrew from the mêlée and was lucky that the Camel pilot did not follow and shoot him down. He was less lucky when he had descended to within a few feet of the ground: a high tension cable barred his path, his machine decelerated abruptly and fell straight down. He suffered a broken nose, cracked jaw, a gash near his eye and burst blood vessels in both legs. The injuries could not have come at a worse time; the war was again entering a period of abundant air activity, during which he should have been able to augment his successes; but it was equally likely that he would have been shot down and killed before he had added even one. The standard of flying, marksmanship and courage in the British squadrons was at least as high as in the German; and in the last six months of the war the RAF's aircraft were better than the *Luftstreikräfte*'s.

Five days passed before Manfred had another success. His conduct was a paradox of maturity and immaturity. Frequently visiting flying training schools to learn who were the most promising pupils, and always aware of the progress of any pilot who began to distinguish himself in action, he sought to recruit the most promising for his *Jagdgeschwader*. Ernst Udet, who had by now destroyed twenty-five British and French aircraft, was one such. Learning that his squadron was now in the vicinity, Manfred went to see him and invited him to transfer to *JG*1; which Udet accepted. Germany was preparing an offensive that would strike at two sectors on the British front and one on the French. As part of his thorough preparation for this, Manfred had even dealt with the detail of nominating his successor, should he be killed, wounded or captured. These were the concerns of a first-rate, adult-minded commander. Yet, when he shot down a Camel on 18 March, his sixty-fifth success, he was so determined to secure his usual souvenir that he followed it down, landed near it and cut out a piece of fabric from the fuselage that bore the serial number: an astonishingly puerile act and all the more surprising because he had ample fuel and ammunition to stay aloft and take part in the fighting or observe the actions of his pilots, on which it was his habit to comment at debriefing.

On 23 March he wrote to tell his mother that he visited Lothar every day and his injuries were healing.

CHAPTER 16

FOOTPRINTS ON
THE SANDS OF TIME

A Psalm of Life – Longfellow

Germany had planned what was intended to be the final assault that would lead to total victory over the Allies. The 17th, 2nd and 18th Armies, comprising seventy-three divisions, would attack the forty-three-mile long Arras–St-Quentin–La Fère Front. The main force was to be exerted north of the Somme. After the 17th and 2nd Armies broke through, they were to wheel north-west and push the British towards the coast, while the Somme river and 18th Army protected the flank.

For some months *JG*1 had been under the 2nd Army, whose area of air operations was now divided into two zones, of which Manfred was in command of the northern one. For this great battle *Jastas* 5 and 46 were added to his strength.

The attack began at 4.30 a.m. on 21 March with a bombardment by 4,000 guns. The thrust broke through south of the Somme but was held up near Arras. Manfred was supposed to lead *Jasta* 11 into the air at 9 a.m., but there was a thick mist that, although helpful to the foot soldiers, did not allow any take-offs until 12.30 p.m. Although 52 sorties were flown, no aeroplanes and only two observation balloons were destroyed.

On 24 March, leading a 25-strong formation against ten SE5As, Manfred increased his victories to sixty-seven. Next day he scored another, on the following day two more, on the day after that another three, and on the 28 March his seventy-fourth.

The early success of the German armies brought unwelcome visitors to *JG*1 when three Members of Parliament arrived to dine, make orotund speeches and stay overnight. The bored and embarrassed pilots wreaked their revenge when the visitors were asleep. With blank cartridges, flare pistols and shouts of 'Air raid' they roused them, saw them run out of their huts in night clothes, and called to them to get back indoors quickly. Early next morning the politicians departed hastily. Manfred enjoyed the comedy as much as anyone.

On 1 April 1918 the Royal Flying Corps and Royal Naval Air Service were combined to form the Royal Air Force. There was no change in uniform but the naval officers were given the equivalent Army ranks e.g. lieutenants became captains and lieutenant-commanders became majors. A year later a new nomenclature was applied, introducing the ranks of pilot officer, flying officer, flight lieutenant (equal to a captain), squadron-leader (major) etc.

The British and Germans were both doing a lot of ground strafing, so combat between aircraft was being carried out at low, as well as medium and high altitudes. Manfred's seventy-fifth victim was an RE8 on 2 April, his seventy-sixth a more worthy Camel on the 6th that had been shooting-up the trenches. His next two were an SE5A at 17,000 ft and a Camel within thirty-five minutes on the following day. Variable weather interfered with both sides' flying programmes on several days, so he had to wait until the twentieth day of the month to get his seventy-ninth, a Camel again. Three minutes later he shot down yet another Camel, his eightieth and last. This was the greatest number of successes achieved in the war. Next came René Fonck with seventy-five and Edward 'Mick' Mannock with seventy-three.

On Sunday 21 April, euphoric with his impressive round-figure total and no doubt looking forward to adding the twenty more that would raise it to three figures, he had to wait for a morning fog to disperse before he could set about it. Soon after half past ten he and four others were airborne with the intention of intercepting enemy aircraft that had been reported coming their way. Another formation of five took off close behind.

Instead of the tail wind from the west that prevailed for most of the year and gave the Germans the advantage, the wind that day was blowing strongly from the east. Presently the *JG*1 pilots saw two flights of five Camels, one higher than the other. There were also five aircraft of *Jasta* 5 in the area, which a third flight of 209 Squadron Camels attacked, while others went for the Triplanes.

Leutnant Hans Joachim Wolff, in Manfred's flight, saw his leader shoot at a Camel that spun out and dived away under control. Before he had to turn his attention to defending himself against an attack from astern that took him by surprise while he was watching Manfred, he saw him still chasing the Camel until both were very low and over the British lines. That was the last that anyone in *JG*1 saw of their Commanding Officer.

All the circumstances of Manfred's death created a conundrum of which several solutions have been put forward and, although the most recent one seems to be supported by the best evidence, it is unlikely to remain unchallenged; and perhaps the truth will never be known.

It was accepted that he had been killed by one bullet. The first attribution of the fatal shot was to a Canadian fighter pilot, Lieutenant Roy Brown of 209 Squadron, which flew Camels, who was immediately seen as the most likely man to have fired it. The sequence of events was that in the general dogfight another Canadian pilot in Brown's flight, who was on his first operational patrol, Second Lieutenant W.R. May, fired such long bursts at various fleeting targets that his guns jammed and he could not clear them; hence his spin away from the brawl. As soon as he levelled out he came under fire from a red Fokker Triplane astern and was so much a novice that he did not recognise it as the dreaded Red Baron's. He was flying so ham-fistedly that Manfred must have been unable to predict his movements and therefore could not get in a crippling shot. The Fokker's maximum speed was 115 mph at sea level and the Camel's was 115 mph at 6,500 ft; so Manfred had a slight advantage.

German ground troops had been firing at May as he and Manfred flew over their trenches. Now, at between sixty and 100 ft over British territory, machine-gun and rifle fire continued, but was aimed at May's pursuer. At last, when May followed a bend in the river and Manfred took a short cut by nipping over a low hill, May realised that he was a sitting duck with his adversary tightly on his tail. But in the next instant Manfred spun into the ground. May saw a Camel close behind his, so formated on it and they flew home together.

This aeroplane was flown by Brown, who had been firing at Manfred and believed that he had shot him down.

There were other claimants. A sergeant gave Manfred a seven-second burst from a Vickers gun as he flashed past and saw him wobble as though hit – so did other witnesses. May and Manfred flew past an Australian artillery battery's position that was guarded against low-flying aeroplanes by two Vickers gunners, R. Buie and W. Evans, who opened up at the Fokker; an Australian rifle platoon also took shots at it and cannot be lightly dismissed. Rifle fire has brought aircraft down on many occasions. The last was when the pilot of an RAF Hunter, a fighter capable of 515 mph, was killed in that way in Yemen during the 1960s when making a ground attack.

Manfred had been seen to move his head and torso suddenly as though he had been hit or were looking round, surprised by coming under fire from astern. He had also been seen to bank. These happenings further confused the argument about who had killed him. Shifting his position could account for the entry point of his fatal wound a second or two later, or it could indicate that he was reacting to sudden pain at the moment of being hit. No post mortem was done, but he was given a hurried medical examination that morning and again twenty-four hours later. The Army and RAF doctors said he had been hit by only one bullet that entered one side of his body and emerged on the other. Even about this there are two versions. According to one, the bullet entered his chest in front near the right shoulder and came out near his left nipple. The other declares that he was hit in the back, near the right shoulder but agrees about the bullet's point of exit. The theory behind both is that it was deflected by the spine on its way through his torso. Whether it penetrated from front or back, Manfred could have been twisting in his seat to look round or banking at that moment.

It is difficult to reconcile one bullet wound with machine-gun fire; and to make deduction even more difficult, later statements by men who viewed the body claimed that they thought they had seen other wounds as well. The argument therefore lies between a machine-gun, either on the ground or in Brown's aeroplane, and a rifle. The mention of several wounds was belated and dubious. A rifle shot that found its mark with or without first ricocheting, or one machine-gun bullet in a burst fired from the ground and deflected into his body by some part of his aeroplane, seem to be the most tenable theories. 'Not proven' suggests itself as the wisest verdict.

The immediate official decision was that it was Brown who had killed Manfred von Richthofen. This was not pronounced by a doctor or doctors but by General Sir Henry Rawlinson, commanding 4th Army, based on the report of medical and other officers who had viewed the body and the crash site and come to this conclusion.

On the next day Manfred was buried with full military honours in the cemetery at the village of Bertangles. The RAF dropped a message to inform his comrades officially of his death and burial. After the war his body was re-interred in a war graves cemetery. In 1925 it was exhumed again and buried with great ceremony in Berlin.

Lothar, who had sworn to avenge his brother, rejoined *Jasta* 11 as Commanding Officer on 19 July. He was shot down and wounded again on 13 August, by when he had 40 victories to his name. He was invalided out of the Service and died as a passenger in a civilian aeroplane accident in 1922.

> Lives of great men all remind us
> We can make our lives sublime,
> And, departing, leave behind us
> Footsteps on the sands of time.

Rittmeister Manfred *Freiherr* von Richthofen exemplified Longfellow's lines. Of the many forms of human greatness, his might not be admitted by pacifists or those who have suffered, or whose families and friends have, from Germany's precipitation of their countries into two great wars within a period of twenty-five years. He was, however, a great patriot, a great leader of men, a great inspiration not only to his fellow airmen but also to the ground troops who saw and admired his bravery. He had the highest possible standard of conduct, integrity in all matters and sense of duty, and was thus a great example to the youth of his country: even though, lamentably, his generation and the next were preternaturally imbued with a belligerent ambition to dominate Europe and, if they could, the world. He was a great professional airman who, although he entered the air war too late to be the greatest innovator, did make a valuable contribution to the development of fighter operations.

His legacy of experience and wisdom was the *Air Combat Operations Manual* he wrote. Its opening words acknowledge his indebtedness to his mentor, Oswald Boelcke: 'Boelcke divided his twelve pilots into two flights'. Manfred went on to explain that each consisted of five or six aircraft and up to six or seven are best led, watched and manoeuvred by their leader.

He dealt first with operations in *Geschwader* strength.

'When the British operate in big formations, or in smaller ones equipped with better aeroplanes than ours, it is necessary to meet them with an appropriately sized force. I favour thirty or forty machines with the leader in front and two squadrons on either side, their leaders 150 metres ahead of them. The two squadrons immediately behind the wing commander fly fifty metres higher than he, with the second two another fifty metres above those.'

A formation of this size should operate only in good weather, aiming to position itself above the enemy and between them and their front line.

A clear briefing on the course to be flown and all other relevant details are essential and as important as the de-briefing on return.

The wing commander's aircraft must be very conspicuously painted and while the squadrons are taking off and forming up he must fly slowly. Each pilot must have a specific position in his flight. The whole formation is to be so disposed that there is room to turn in any direction and the wing leader's movements are to be followed immediately (i.e. a dive, climb or turn). Unnecessary about-turns must be avoided because it takes so long to resume formation.

Every pilot must count the enemy aeroplanes as soon as

sighted. Before an attack the wing leader must regulate his speed so that any pilots who have straggled can re-form.

Since the purpose is to destroy enemy formations, the wing commander should not attack individual aircraft, nor should he concern himself with enemy machines that break formation, but leave them to the squadrons astern. The pilot nearest to an enemy is the one who must engage it. Any pilot who loses altitude during a fight must climb back as soon as possible.

Manfred defines what he expects of flight and squadron commanders. First, he puts comradeship, strict discipline and the necessity for every pilot to have complete trust in his leader.

Pilots must develop diversity by not always flying in the same position. Leaders must adjust their speed to ensure that the slowest aircraft can hold station.

Dealing with attacks, he differentiates between the tactics to use against single aircraft and formations. Describing his method when leading a squadron against artillery spotters, he says that he keeps several possible targets in view, noting their altitudes and changes of course and whether or not they have fighter cover. He then withdraws to gain height and returns towards the enemy front with the target he has selected below him. Allowing for wind direction, he leads the formation in a dive with the sun behind them onto the enemy machine. He says that to leave any aircraft at a higher altitude as cover is cowardice. Whoever is first in position to shoot at the target has the right to fire first and two pilots must never fire simultaneously. If the enemy pilot sees his attackers, he will usually dive away or stay, at a low altitude, turning. Attacking a turning target is seldom successful, so it is best to regain height and repeat the attack on the same or another machine at a greater height.

He emphasises that whatever the size of a formation, it is essential that its leader never loses sight of the whole battle; a breadth of vision and attention that can be attained only by fighting in several wing combats, and which is the most important attribute.

About training novices, he says that 'the main thing' for a fighter pilot is his machine-gun. He must be able to recognise the cause of a jam. It is his own responsibility, not the armourer's, to ensure the perfect performance of his gun. When a gun jams, Manfred blames the pilot and says that a machine-gun in good working order is more important than a smoothly running engine. When loading ammunition belts a pilot must make sure that every bullet is measured precisely.

He expresses the low value he places on flying ability and justifies this by pointing out that he scored his first twenty victories while still finding great difficulty in aircraft handling. He rates a pilot who can turn only in one direction but attacks boldly, higher than one who can throw his machine about the sky impeccably but attacks with caution. The need, he insists, is for daredevils, not aerobatic artists.

Formation practice at high altitude, including tight turns at full throttle, must be done; so must long-distance navigation exercises so that pilots learn to know the terrain well enough not to need a map.

Prescribing tactics to use against single- and two-seater aeroplanes, he says that fighting a single-seater is much the easier. The first essential is to apply Boelcke's advice about shooting only when within fifty metres of the target. The others are to take the pilot by surprise, make tighter turns than he and keep above him. When himself attacked from above, he makes all dives and turns at full throttle and tries to climb while turning, to gain height over his attacker.

Two-seaters must be attacked from astern at high speed and the observer must be killed or incapacitated by wounds. Shots are most effective when the target is flying straight or in a turn, but tracer can be fired from the beam or when it is banking, to scare its pilot. It is most dangerous to attack a two-seater from ahead, as you are then within the field of fire of both front and rear guns; and when turning astern to attack again, you are fully exposed to the observer's fire. If a two-seater attacks you head-on, slip under it and start a tight turn as it passes overhead.

His final admonition is the one he ignored, fatally and for the first time, in his last fight. 'You should never stay with an opponent whom, through your bad shooting or his skilful turning, you have been unable to shoot down, the combat lasts for a long time and you are alone, outnumbered by adversaries.'

CHAPTER 17

THE GREAT ALLIED FIGHTER LEADERS IN 1914–1918

T he first commander of a fighter squadron to be rated among the greatest in the history of the RAF was Lanoe Hawker. The outstanding precursor in any activity has an advantage over his successors, because he is the first to set a standard and become conspicuous. His achievements might not in themselves be superlative, but by comparison with those of his contemporaries they are the best.

A great fighter leader has to be judged on several qualities and attainments. It is difficult enough to attempt comparisons with famous sportsmen and women of different generations, where only one attribute is in question. Would the World Heavyweight Boxing Champion of 1910 have beaten the champion of 1990? Would the winners of the men's and women's Singles at Wimbledon in 1939 have beaten those of 1969? Would the rowing VIII that won the Grand Challenge Cup at Henley in 1900 have beaten the 1995 winners? It is impossible to say, even by feeding all the relevant data into a computer. A fighter leader has to be assessed not only on the number of his victories but also the quality of the aircraft he and his adversaries fly, the ground support available, the innovations he introduces to fighter operations, the influence he exerts over those he commands and his effect on the development of air fighting.

The obvious essentials that the best fighter leaders have are courage, intelligence, a strong personality and an understanding of human nature. They must be able to get the best possible performance out of their pilots as well as themselves excelling. The common factor shared by all great military leaders in any Service is the over-worked noun 'charisma' in its modern sense: the outstanding quality in a person that gives him or her influence and authority over others.

Hawker shot down only nine enemy aircraft, but he did so in the most difficult circumstances: at first flying primitive aeroplanes with makeshift armament and later a purpose-designed fighter that was still poorly armed and soon out-performed by enemy fighters.

In addition to being brave and clever he had a cheerful nature that made light of difficulties, a fine example to others, and was popular as well as respected and admired. He was known as 'jolly old Hawker', with the emphasis on the first adjective. His first squadron, No. 6, with which he went to France in October 1914, was equipped with two Henri Farmans, three BE8s, and five BE2s. Reconnaissance and artillery spotting were its functions. He flew one of the Farmans, which were the oldest of these three types, for a month until it was destroyed in a gale. During that time he put in 46½ flying hours, more than anyone else. He was next given an old BE2, which another gale wrecked six weeks later. This was replaced by a new machine of the same type. He always carried three 20 lb bombs to drop on any suitable target and whenever he saw an enemy aircraft he attacked it with revolver or rifle fire.

From the first he showed a quality that he shared with Manfred: admiration of, and sympathy for, the infantry, for whom he would fly in any weather and take any risk.

In April 1915 he was sent to bomb a zeppelin shed over which an observation balloon was kept, surrounded by anti-aircraft guns. He dropped two bombs from 4,000 ft, then, after dropping hand grenades at the observer in the basket beneath the balloon, he spiralled down around the mooring cable, dropped his last bomb from 200 ft and hit the target. Several bullets perforated the BE2. For this he was awarded the Distinguished Service Order.

In June 1915 he began flying a Bristol Scout, to which he had a mounting fitted for a Lewis gun beside the cockpit. In July

1915 he became the first RFC or RNAS pilot to shoot down an aircraft in flames: not using incendiary bullets, which the RFC spurned; the petrol vapour ignited spontaneously. In the same fight he forced another down to a crash-landing. In twelve sorties he engaged a total of fifteen enemy aircraft, brought five down and put nine to flight. In August he was awarded the Victoria Cross.

The spirit with which he later imbued No. 24 Squadron when he became its first commander, and the fight in which Manfred von Richthofen killed him, have already been described. Given an aeroplane that performed as well as Manfred's, he would most probably have been the victor for he was a far superior pilot and a great deal more intelligent and subtle. His influence over fighter operations was profound: he displayed unrelenting aggression and was unrivalled as an innovator, thanks to his training as an engineer. Among these were the double ammunition drum and a device for holding the gun on a DH2 steady when fired instead of wobbling. It was he who had the first fur-lined thigh boots made, which became standard issue. He never became a hater of the Germans. When they earned worldwide condemnation for executing Nurse Cavell for helping British prisoners of war in Belgium to escape, he accepted that they were behaving according to their notion of what was right and it was the duty of every Briton to do his utmost to ensure that their own country never fell under such an enemy's cruel domination.

Ball stands high among those who had an enduring influence over his own and future generations of fighter pilots because of his reckless bravery and the tactic he introduced of attacking from close astern of, and beneath, his chosen victim. His youthful attitude to a deadly serious adult occupation captured everyone's imagination. His comrades knew him as meticulous about every detail that would improve the performance of his aircraft and his gun. He was probably the first RFC pilot always to load his ammunition drums and belts himself. He was not gregarious. If he could not have a bedroom to himself he preferred sleeping in a tent to sharing one in the standard wooden, or iron Nissen huts, and entertained himself by playing the violin in his quarters rather than being sociable in the mess.

Mannock, as the highest-scoring British pilot, of course had a strong influence over the doctrine of fighter operations. His great concern for those whom he led, the trouble he took to teach them how best to fight and to protect themselves were an object lesson for future squadron commanders to emulate. If he had had a good education at school and gone to a university he would probably have risen to Air rank – had he survived the war and stayed in the Service. The most emotional of all fighter pilots, he was an intellectual with an enquiring mind, singleness of purpose and a great capacity for hatred of anyone whom he perceived as being in any way an enemy.

McCudden was the great exemplar of the professional devoted to the military life, took naturally to flying, accepted authority and discipline and knew how to exercise both. His natural leadership was enhanced by the years he spent learning from the leadership of others as he advanced from the lowest rank to that of major. Like Hawker and Mannock he took pains to teach inexperienced pilots how to fight and stay alive. His fatal accident that resulted from a disregard for rules was an inexplicable aberration. He too, had he survived the war, could have risen to the rank of air marshal.

The second-highest scoring RFC fighter pilot, with only one less victory than Mannock, W.A. 'Billy' Bishop, was not, despite his VC, DSO, DFC, a great leader; as Mannock discovered when he took over the demoralised pilots of 85 Squadron from him. In 1911 he had entered the Canadian Royal Military Academy, from which his escapades courted expulsion. His independent attitude remained undiminished when he had earned his observer's wing and arrived at the Western Front in January 1916, after fifteen months in the cavalry. Following a crash in which he injured a knee, he passed the pilots' course and spent some months in a home defence squadron, joining 60 Squadron in France on 7 March 1916 to fly the Nieuport

17. He scored his first victory, over an Albatros, on 18 May and only two days later led a section. By the end of April he had shot down twelve German aeroplanes and been promoted to captain.

He hated Germans for having started the war, despised them and thoroughly enjoyed killing them. On the excuse that he did not like to have others' lives in his hands, he used to go off alone as often as possible. When he saw British infantry being cut down by machine-guns it aroused his fury. On one occasion he dived vertically, shooting at a group of the enemy who were manning two Spandaus in the corner of a trench. One of these guns returned his fire. From a height of thirty feet he 'could make out every detail of the Huns' frightened faces. With hate in my heart I fired every bullet I could at them'. Immediately after, he saw the British troops advancing again.

'Soon after this, new Hun-hatred had become part of my soul', he wrote, he attacked a two-seater that turned tail and landed. To see the German alight under perfect control 'filled me with a towering rage. I vowed an eternal vendetta against all the Hun two-seaters in the world'. He dived to within a few feet of the ground and fired at the machine. 'I had the satisfaction of knowing that the pilot and observer must have been hit or nearly scared to death'. He waited a while but there was no sign of life in the riddled aeroplane.

In May he won the MC, in June the DSO and in August the VC.

The third highest scoring pilot has not been written about as generously as he deserved. He was another Canadian, Raymond Collishaw, and his comparative lack of recognition is owed to the fact that he did not serve in the RFC, but in the Royal Naval Air Service, from 1915 until the RAF was formed. His rank of lieutenant-commander became its equivalent, major; when the newly created RAF ranks were announced after the war this changed to squadron leader. Not only did he shoot down sixty-eight German aeroplanes, but also his was an outstandingly forceful personality, even among others who were conspicuous for this quality. To those who served under him in both world wars he was the epitome of what people mean when they describe someone as 'a character'.

There were so many delays in sending the first draft of RNAS trainee pilots from Canada to Britain that 'Collie' Collishaw did not arrive at the Western Front until 1 August 1916. Had he gone into action sooner he might well have surpassed not only Mannock's and McCudden's scores but Fonck's also, for he survived to become the first commander of Desert Air Force in 1940 and retired as an air vice-marshal. Of his first operational sortie, he said 'Anyone who claims not to have been nervous on such an occasion has to be an insensitive idiot or have a bad memory'.

In 1917, fighting an Albatros D at 16,000 ft, he was turning his Sopwith Triplane inside his opponent and watching another D that was trying to get on his tail, when a third Albatros loomed dead ahead. He dived steeply and barely scraped beneath it. The violence of the manoeuvre broke his safety strap and he found himself hanging outside the cockpit, holding the two centre struts. Suddenly the Tripe's nose reared up, then it stalled into a spin. A strong man, he managed to hold on. The aeroplane pulled up sharply again and he was able to hook a foot round the joystick, put the machine into level flight and tumble back into his seat. By then he was down to 6,000 ft.

It was probably Collishaw, then a flight commander in No. 10 RNAS Squadron, who nearly killed Manfred von Richthofen by wounding him in the head. His flight was escorting twelve FE2s and RE8s when *Jasta* 11 attacked. Manfred himself naturally went for a Fee – the Sopwiths would have been too lethal to tackle. Collishaw fired at him, but it is impossible to be certain that it was he who hit him.

Among the French, the first to show the characteristics of a natural fighter leader was *Capitaine* Félix Happe, who commanded a bomber squadron, MF 29, in GB 4, one of the first four bomber groups. He was a colourful character, over six feet tall, with a bushy black beard parted in the centre and beetling eyebrows – and a great sense of humour. It was he who initiated the V formation that became standard in both the Allied and enemy air forces. On bomber raids he flew his squadron at a greater altitude than the others, to give them high cover before bombing, even when they were escorted by Nieuports. On the way out and back his pilots did not merely keep a look out, they actively searched for the enemy. After his squadron bombed a poison gas plant and the Aviatik works on the same day, 26 August 1915, the Germans put a price of 25,000 marks on his head. He sent them a message to say that, to avoid wasting time on anyone else, they should note that his aeroplane was recognisable by its red wheels.

In 1916, promoted to *Commandant*, the equivalent of major, he took command of BG 4. At the suggestion of the French, to which the British Admiralty agreed, an Allied bombing squadron was formed to attack German munitions factories. As well as bombers, the RNAS provided fighters to escort them, among which was Collishaw's flight.

The first French fighter leaders to distinguish themselves all held the rank of *capitaine* and commanded Morane Saulnier squadrons. The earliest was Félix Brocard of MS 3, followed by Tricornot de Rose of MS 12 and de Vergette of MS 23. It was now that, having attained greater distinction than any other fighter *escadrille*, MS 3 adopted the stork as its symbol and became known as *Les Cigognes*. Presently it was re-equipped with the Nieuport 11, so became N 3. By the end of 1916 it had grown into a *Groupe* which also comprised N 26, N 23, N 73, N 103 and N 167, still under Brocard, now promoted to *Commandant*.

Fonck was an inspiration to his comrades rather than one who displayed the eminent leadership qualities of the outstanding British and Empire fighter leaders. Brilliant individuals abounded and their successes were a tribute to bravery and ability, but even he, who topped the Allied list of victories, was renowned for his skill and the advice he gave, not for flying at the head of a formation or originating tactics.

The same is true of the Italians, except that their most successful pilots did not achieve fame outside their own country and were known only to those RFC pilots whose squadrons fought on the Italian Front.

*   *   *

The victory statistics show the disparities in individual performance.

Number of pilots with

	70–80	60–69	50–59	40–49	30–39
RFC and RNAS	2	1	5	5	14
French	1		1	2	1
Italian					1

CHAPTER 18

THE GREAT FIGHTER LEADERS IN 1939–1945

I n the Second World War the gift of leadership was abundantly apparent in the RAF, including the squadrons consisting of pilots and ground crew of various Commonwealth and European nationalities. This talent was also conspicuous among the *Luftwaffe*'s *Staffel* and *Geschwader* commanders. It is invidious to evaluate outstanding fighter leaders on the basis of the number of victories they scored; not only because opportunities varied between the various campaigns but also because the number of enemy aircraft a pilot shoots down is neither the only nor the most important criterion. What distinguishes a first-class leader from those less gifted is the overall grasp of a situation, the positioning of his force to the best advantage, the ability to predict the enemy's tactics, to follow the course of a battle and to recall it afterwards at de-briefing. This is what is known today as situational awareness.

The first of the most obvious differences between the performance of fighter pilots in the two wars is in the individual totals of aircraft shot down. In comparison with Mannock's seventy-three victories, the RAF's top-scorer, M.T. St J Pattle, was credited with forty-two. A South African who joined the Service in 1936, he operated in North Africa and Greece against predominantly Italian aircraft. During the last days of the Greek campaign some of the squadron's records were destroyed and he was killed. It is estimated that his real total was around fifty. The RAF's next most successful pilot was J.E. 'Johnnie'

Johnson, with a confirmed score of thirty-eight; all were of high quality, one a Messerschmitt 110 and the rest Me 109s and FW 190s. In terms of difficulty, therefore, he ranks the higher, since many of Pattle's kills were against considerably weaker opposition.

The second glaring disparity is the vast numerical gap between the most successful RAF and German performers in the Second World War. Manfred's eighty victories seem paltry in contrast with the leading aces of the next generation. The supreme German champion, Erich Hartmann, shot down 352 aircraft. Seven were United States Army Air Force P-51 Mustangs. All the others were Russian, mostly much inferior to his and flown by poorly trained pilots whom he might have hacked down in the same astronomical numbers if they had been flying the latest Messerschmitt. Altogether 106 *Luftwaffe* pilots scored 100 or over. Most of these fought the Russian Air Force, some throughout their war service, others during part of it. The RAF never fought any unskilled and poorly equipped enemy except in the early days of the North African campaign and in Greece, where some of the Italian aeroplanes they knocked out were second-rate, even though the pilots had been well trained.

There is a significant and supremely relevant statistic common to both that is at variance with the general conception of air fighting. For the public at large, the assumption about fighter operations, particularly in the Second World War, when so much of it was carried on within the view of the population of England, Wales and Scotland, is of British pilots each shooting down German aeroplanes in large numbers. The exploits of a small number of the most successful among these, with victories in double figures, were reported by newspapers, on the radio and in cinema news reels. The general assumption was of a fairly even spread of personal victories over the six years of the war. The truth is that in both great wars, approximately five per cent of pilots shot down approximately forty per cent of the total bag. This, as Mike Spick has pointed out in *The Ace Factor*, is attributable to the higher degree of situational awareness (SA) with which this small number of star performers were blessed.

It is well known that good eyesight is mandatory for Service aircrew. Nearly all outstanding fighter pilots have had exceptionally acute long-distance vision. This not only enables them to spot enemy aeroplanes before anyone else in their formation, but also contributes to their SA. The major exception was Mannock, with one good eye and one that was purblind.

Luck plays a part as well as natural and acquired skills. Countless pilots have flown scores of sorties on which they did not see the enemy at all or at too great a distance or height to permit an attack. No amount of natural SA can make up for lack of opportunity.

The RAF's best fighter leaders all had high personal scores – that was why they were given command of squadrons and wings – but it is the doctrine they presented to their pilots and their own situational awareness that are most significant.

Every squadron commander drummed into his pilots the supreme importance of attacking from up-sun, from a greater height than the enemy, and of opening fire only when as close to the target as possible. When flying in pairs or multiples of pairs became standard in both the RAF and the *Luftwaffe* there was another rule to observe; the No. 2 must stay in position and follow his leader's every move, always covering him from a surprise attack, warning him when to break if that were the only resort. Even the naturally brilliant shots, such as James 'Ginger' Lacey, the RAF's top-scoring pilot in the Battle of Britain, 'Pat' Pattle, Adolf Gysbert Malan, Robert Tuck, 'Johnnie' Johnson, George 'Screwball' Beurling, Frank Carey, Douglas Bader, Colin Grey and others closed the range as much as they could rather than risk both missing the enemy aeroplane and wasting ammunition.

At the beginning of the war, Hurricanes' and Spitfires' eight .303-inch Browning machine-guns were harmonised to converge at a distance of 450 yards, known as 'the Dowding spread'. When Fighter Command first went into action its pilots found that they could not inflict enough damage to bring an enemy aeroplane down or even, sometimes, to hit it at all. The favoured harmonisation soon became 300 or 250 yards. 'Ginger' opted for 150; even though he was a fine shot, he believed in almost breathing down his adversary's neck before he pressed the firing button.

Douglas Bader is remembered all over the world because, despite both his legs having been amputated, he shot down at least twenty-three enemy aircraft and probably seven more. The RAF remembers and reveres him for his impeccable aircraft handling, brilliant fighting and his bravery, which although equalled was never excelled. Those who flew in Fighter Command during the Second World War mostly assess him the most inspirational fighter leader of the whole war. It was not only his airmanship, accurate shooting, dauntless spirit and authoritative personality that raised him to such heights, but also his high degree of situational awareness. He was also an innovator; the first in the RAF to adopt the finger-four formation and to advocate attacking enemy raids during the Battle of Britain in Wing strength. The argument about the merits of sending up three squadrons together instead of one at a time persists today despite the fact that there are few still alive to discuss it from personal experience.

Douglas was misunderstood and misquoted at the time and has been ever since. The situation was that fighters in 11 Group of Fighter Command, which covered southern and south-east England, were the first to be ordered up when an air raid was reported approaching Kent or Sussex. The biggest formation was squadron strength. Often, by the time they had climbed high enough to intercept the raid it had already dropped its bombs. Douglas, who was in 12 Group then, which was immediately to the north of 11 Group, suggested that 12 Group should be scrambled as well, and in a Wing of three squadrons, not squadron strength, which was too small to be fully effective.

Those who disagreed with this contended that to send a 12 Group Wing up when 11 Group was heavily outnumbered would be too late. The whole point of Douglas's plan was to scramble them at the same time as 11 Group, so that they would be able to make altitude comfortably. If the radar warning was tardy, then 11 Group could be dispensed with, because they would not be able to climb high enough in the time available, and No. 12 alone sent to intercept the enemy. His opponents appeared to be deaf to this. The contra-argument did have a certain validity, namely that handling thirty-six aircraft was a cumbersome business, starting with the collision risk as they milled around forming up. Soon after the Battle of Britain, sweeps over France by Wings of three, four or five squadrons, escorting a dozen bombers, were frequent, flexible for the Wing leader to handle and collisions rare. In fact, those who argued against Douglas heard only what they wanted to and affected not to have heard what he was actually saying. In any event, they lacked his intelligence and resented anyone of his rank propounding a procedure to them, who were one, two or more ranks senior.

Underlying the opposing theories was another pungent complication; the *amour propre* of the air vice-marshals commanding the two groups. One did not want intrusion in his patch of sky. The other was eager to spread his sphere of influence. Those are the familiar irreconcilable factors in most joint endeavours.

The same tributes to intelligence, courage and personality that have been paid to Douglas can be applied to 'Johnnie' Johnson; who did not score his first victory until 26 June 1941, by which time Douglas, who had first tasted blood on 1 June 1940, had fifteen on his score card. As leader of a Canadian Wing based at Kenley and later of another Canadian Wing which, as part of 2nd Tactical Air Force, played a prominent part in the invasion of France in 1944 and all the fighting that ensued until the armistice, he shone as one of history's finest fighter commanders in every respect. His DSO and two bars and DFC and bar are evidence of that. In 1950 he served in Korea on attachment to the United States Air Force, which awarded him the US Air Medal and Legion of Merit.

The Germans learned the same rules and in the same way. As for gunnery, the Me 109E was armed with two MG17 7.92mm machine-guns with 1,000 rounds each above the engine, or one MG17 with 500 rounds and, firing through the propeller hub, a MG FF 20mm cannon. Both marks also had two MG FFs in the wings. The destructive power of a cannon shell was formidable and effective at a greater range than a bullet.

There was another important factor, the application of one of the facets of situational awareness; when possible, to take time selecting a target so as to catch the enemy by surprise instead of charging into a fight without first making sure your intended victim was not the bait in a trap. Manfred had been cautious in this way, but Lothar never. Even at the instant when a formation, not just a single aircraft, was about to engage another, there could be time to make a reasoned choice.

The first masterly fighter leader and tactician in the *Luftwaffe* was Werner Mölders. During the Spanish Civil War that began in 1936 and continued into 1939, Germany and Italy each sent an air component to fight for the Fascists; the former's was known as the Condor Legion. The USSR sent air support for the Communists. Mölders commanded a Me 109 squadron, shot down fourteen hostiles and gained valuable experience. He is credited with having invented the finger-four formation of two pairs of fighters in line abreast, which he found much more efficient than a V of three or diamond of four.

However, another German pilot in the Second World War, *Oberleutnant* Otto Stammberger, Stotto to his friends, who flew Me 109s and FW 190s, shot down a Spitfire and six B-17s and was wounded, says,

> 'The great innovator and leader in tactics was Oswald Bölcke. He propagated the finger-four formation, with the accompanying aircraft on his flank. But this idea was not carried straight through to WW2; the German fighter pilots in Spain snatched it up and developed it as the "four flight", the ideal formation.'

This attribution of the origin of the finger four to Böelcke appears to have been generally forgotten outside Germany, although Johnny Johnson mentions it in *The Story Of Air Fighting*; but Mölders is a worthy inheritor of the credit for having originated the bright idea, even if it is not entirely merited.

Stotto continues, 'In our day, Richthofen was idolised, but

today's youth have forgotten him!' He also says 'Richthofen had the best chance to shoot down enemy aircraft. As leader of a formation, he saw them first, was closest and attacked first.' It is implicit that he was also protected by the others; and if the formation chanced on a singleton, it was he who attacked it.

Mölders was the first to surpass Manfred's total and the first to score a century of victories – including those in Spain and seventeen French fighters and bombers. His talent for leadership, organisation and as a tactician and teacher was outstanding. He was conscious of the traumatic effect on a novice of being badly frightened in action, by being 'jumped' by the enemy or seeing his comrades go down, perhaps in flames. One of his maxims was 'It is most important for a young fighter pilot to gain his first victory without undergoing a great shock.'

He became General of Fighters in 1939 at the age of twenty-eight with his score at 115. On 22 November 1941 he was killed when the Heinkel 111 in which he was a passenger crashed.

He was succeeded as Commander of the German fighter force by Adolf Galland who, happily, survived the war, in which he reached the rank of Lieutenant-General, and has been described as by far the best known German fighter pilot and leader of the Second World War.

Although Galland also went to Spain with the Condor Legion, he was disappointed by having to fly Heinkel 51 biplanes, superannuated fighters that carried four 50 kg bombs and were being used for ground attack, in which he flew 300 sorties. His report on this experience so impressed the German Air Staff that he was given command of a squadron of Heinkel 123 dive bombers. Determine to fly modern fighters, he enlisted the collusion of a medical officer, complained of deafness caused by flying in an open cockpit, and soon found himself snug in the enclosed cockpit of an Me 109E.

Clever, with a receptive mind, brave and conscientious, the burden of high office was imposed on him at the age of twenty-nine. Between 12 May 1940 and 8 November 1941 he scored 103 victories, which included five French Air Force machines and three Belgian. He ended the war still fighting.

Having antagonised both Göring and Hitler, he was relieved of his appointment but kept his rank and, in January 1945, took command of Germany's second jet squadron, JV44, flying the Me 262. Among the lieutenant-general's pilots were two colonels, a lieutenant-colonel, three majors and five captains. The sixteen others were lieutenants and second-lieutenants. In April 1945, he fired a salvo of 24 rockets at 16 USAAF Marauders from 600 yards and destroyed one. On 26 April a Mustang took him by surprise, wounded him and sent his aircraft down. For a fleeting instant his SA must have lapsed.

The most successful American fighter pilot was Richard Bong, with forty victories in the Far East Theatre of Operations. Close to his score came Thomas McGuire with thirty-eight and David McCampbell with thirty-four. In Europe, the highest USAAF scorers were Frances Gabreski with thirty-one, Robert Johnson with twenty-eight and George Preddy with twenty-six.

Dick Bong modestly used to say that he was a poor shot when he first flew in action but improved with experience. This is true of many with high scores. The opportunities for his compatriot fighter pilots operating in Europe were limited by their comparatively short operational tours of 200 hours and the nature of their duty, which was to escort bombers on daylight raids that entailed very long flights and therefore relatively few sorties. Further, their orders during the first many months were to stay in formation and not break away to initiate attacks when they spotted enemy fighters.

Wing Commander Lance Wade DSO, DFC, an American who began flying in 1933 and owned his own aeroplane, must be mentioned here. He joined the RAF in 1940, shot down twenty-five German and Italian aircraft in North Africa and Italy, and was killed in 1944 in an air accident.

Again, it was a high degree of situational awareness as well as marksmanship that distinguished these men.

CHAPTER 19

THE KOREAN WAR

In June 1950, communist North Korea invaded democratically governed South Korea and began a war that lasted for three years. China provided the aggressors with USSR-built fighter aircraft, mainly MiG 15s, which were flown by Russians and Chinese as well as Koreans. To help the South, the USA sent fighter squadrons equipped with the F-80, F-84 and F-86; also piston-engined P-51 Mustangs for ground attack. Australia contributed a Meteor 8 squadron, but the air superiority war was to be between fast jets and this early type was too slow, so was also engaged in ground support. RAF Sunderland flying boats patrolled the coasts and open sea. RAF night fighting experts served as advisers on night intruder operations.

Essentially, fighter-to-fighter combat was still the same as it had been in 1914–1918: the aeroplanes were armed with guns, not homing missiles, so the pilots had to estimate the amount of deflection to allow, fly accurately and foil their opponent's aim by evasive action. They also had to spot and identify their targets visually, not on a radar display, as has been possible for day fighters in later wars. There was another link with the Second World War: several of the American pilots had fought in it. Among them was Frank Gabreski, with twenty-eight victories in Europe already to his name, who gained six and a half more. The top ace was Joseph McConnell Jr, with sixteen. Next came James Jabara, who made fifteen kills.

The main difference between air fighting in 1939–1945 and 1950–1953 was that both speed and rate of climb had increased by some fifty per cent. Obviously, this meant that mental and physical reactions and actions had to be faster than ever before.

It also meant that the effect this had on radius of turn seldom allowed more than one shot at the enemy, as well as lost opportunities of a kind that seldom occurred in either of the Great Wars. An instance of the latter happened early on when four F-86s flying north suddenly saw twelve MiGs flying south below them: it was too late to turn and get within shooting range. Another difficulty was the amount of punishment that big, strong jets could suffer without being brought down. The original 50mm guns of the F-86 were the same as those in American fighters five years earlier. Eventually, these were replaced by 20mm cannon.

Whereas both World Wars had seen the growth of fighter formations in battle increase from one squadron to Wings comprising three, four or five squadrons, operations over Korea saw a diminution in formation size. The enemy continued to fly in twelves, but the basic USAF unit was a pair and two pairs usually operated together, but not exactly in the old finger-four. The first element of two was ahead of and 1,000 ft below the second pair, No. 1 leading with No. 2 100 yards astern and to one side, guarding the rear. No. 2 seldom fired his guns unless so ordered by the leader or was separated from the others. The second pair were similarly positioned in relation to each other. When they entered the area in which they expected to meet the enemy, they all adopted the new fluid-four, which meant that the upper pair was higher above the lower pair and more widely spread than on the approach.

The advances in aircraft performance meant that combat height increased. The greatest altitude at which a fighter pilot fired his guns in World War Two was 44,000 ft. This unique event occurred on 12 September 1942 when Pilot Officer Prince Emanuel Galitzine, flying a Spitfire IX of the High Altitude Flight, which was stationed at Northolt, intercepted a Junkers Ju 86P high-altitude reconnaissance type. He was at 38,000 ft when the ground controller guiding him towards the target informed him that the 'Bandit' was at 42,000 ft. Galitzine's combat report records that he could see its black trails. Both aircraft climbed to 43,000 ft. He got above it, dived, opened fire and hit its starboard wing, after which one gun jammed.

The Junkers made evasive turns, Galitzine attacked again, but its slipstream obscured his windscreen. This happened twice and finally the target disappeared in a large patch of mist. It got back to Germany and the two pilots met after the war.

In Korea, dog fighting at 40,000 ft was commonplace. At such altitude the ability to discern other aircraft deteriorated greatly. Recognition also became increasingly difficult, for the look of the various types of jet fighter did not differ from one another as much as piston-engine types used to.

The division of victories between fighter pilots remained much the same as in the two world wars: approximately five per cent of them shot down about forty per cent of their side's total bag.

Mike Spick points out that 'Part of situational awareness consists of knowing the strengths and limitations of the air-craft/weapons systems employed'. With the passage of time, therefore, the demands on fighter pilots had steadily increased and the first appearance of every new enemy fighter type entailed study and experience. Since the Korean War, electronic equipment in the cockpit has eased the fighter pilot's span of attention as dramatically as the installation of airborne radar in two-man night fighters did during the winter of 1940–41.

Finally, two inventions that would have astonished Richthofen marked the end of an era of fighter operations and the beginning of a new one. The first was the helicopter and its effectiveness for ground strafing; the other was the homing rocket, first fired by the F-86H towards the end of the conflict.

CHAPTER 20

CONCLUSIONS

Whant were Manfred von Richthofen's strengths and weaknesses; what good and bad qualities did he possess as a person, a pilot and a leader; what influence did he have over fighter tactics and the use of fighter aircraft; how does he compare in ability with other famous fighter leaders; why is his name so well-known when other eminent fighter commanders have been forgotten?

More has probably been written about him than any other military pilot and he has been the subject of frequent research since the end of the Great War. He wrote a brief autobiography in 1917 and many of his letters home reveal feelings that he would not have expressed elsewhere. His fame exceeds any of his contemporaries' and successors' for two main reasons; not only because of publicity in his lifetime or because he shot down more enemy aircraft than any other pilot of any nationality during his war, but also because no other flying man has ever had so arresting a sobriquet as The Red Baron or led a formation with so striking a one as The Flying Circus.

It is remarkable that his reputation is as great in the English-speaking world as it is in Germany; except, it seems, among today's youth. People may not know the details of his career or even that he was the supreme ace of the Great War, but they have heard of The Red Baron and know that his name was von Richthofen. Ask them to name an outstanding British or French pilot of the same war and few would be able to. Refer to the Second World War and Johnnie Johnson, Sailor Malan, Ginger Lacey, Hartmann, Marseille or Galland would evoke the same negative response. Douglas Bader is probably the one universally well-known pilot of the 1939–1945 war because he flew with two artificial legs, was the subject of more publicity than any other RAF pilot during the war and after it until his death in 1982; and is still sometimes mentioned in the press, often in connection with the Douglas Bader Foundation, which helps limbless people in many ways.

The first characteristic misapplied to Manfred that must be disabused is fearlessness. Raymond Collishaw's words on that subject have already been quoted. Manfred, like all airmen, sailors or soldiers who have been in action and not run away, all firemen and firewomen, doctors, nurses or ambulance crews, doing their duty in an air raid; lifeboat crews and rescue teams in mining disasters and earthquakes were not fearless; they were, and are, brave.

Perhaps the one literally fearless pilot in the Second World War was Flight Lieutenant Richard Playne Stevens DSO, DFC of 151 and 253 Squadrons. A professional pilot before the war, with 400 hours' night flying between London and Paris carrying a cargo of newspapers, he was thirty-two years old when war was declared. This was the maximum age for eligibility as an RAF pilot. He joined immediately and by the end of 1940 was engaged solely on night sorties in a Hurricane. After his wife and children were killed in one of the first raids on Manchester, he showed what the word 'fearless' really means. Consumed by hatred for the Germans, he flew with total disregard for his life. He shot down enemy bombers from such close range that once, when one exploded, his own aeroplane was spattered with the flesh and blood of the men he had just killed. He refused to allow this defilement to be washed off. He had scored his first victory on 15 January 1941 and enjoyed his fourteenth and last on 3 July the same year. His total exceeded that of any pilot and observer who had the advantage of flying in radar-equipped night fighters at that time. On 12 December, as everyone expected must happen eventually, he failed to return from a sortie. Perhaps it is only when, in the words of many coroners' verdicts, 'the balance of his mind was disturbed', that any fighter pilot has ever literally shown fearlessness as distinct from courage.

Manfred was energetic, a characteristic he shared with many

RFC contemporaries who were addicted to games. In the Second World War this was not so evident and the time came when RAF aircrew were subjected to two hours' compulsory exercise every week. He wrote to his mother 'At last I have found an outlet for my energies. On the days free from trench duty, I go hunting'. When he was able to expend his energy on flying he no longer felt restless and his shooting forays on leave were for relaxation: which, for him, seemed essentially to entail slaughter.

Pilots are trained to be painstaking and thorough, which with him was intrinsic. Remember his description of his first flight: 'The night before, I had gone to bed earlier than usual to be fresh for the great moment next morning'. Whatever he undertook he did with enthusiasm and the determination to master it – a born winner. One has only to recall his equestrian exploits for proof of this and of his sheer guts. Before the great event at Breslau: on the day before his mare was put on the train, 'I could not resist taking her over the hurdles in our training area once more. In so doing we slipped ... I cracked my collarbone'. It did not prevent his competing and doing well. Next year, in the Kaiser Prize Race:

> 'I galloped over the heather and then suddenly landed on my head. The horse had stepped into a rabbit hole and in the fall I had broken my collarbone. I remounted and rode another seventy kilometres with the injury'.

In contrast, he showed caution in battle – until his last, when, certain of an eighty-first kill with an obvious tyro as victim, he was abandoned by those who ought to have protected him from an astern attack.

Such affection as he had for animals arose from admiration for their handsome looks and their usefulness to him; they served him. His dog was a companion; his horses enabled him to distinguish himself in competition. One does not need to be a professional psychologist to divine why so small a man chose such a huge pet.

He was cruel. He made unreasonable demands of the horses he rode, he thrashed his dog, he exulted in killing game, big and small, and in sending his victims in the air to their death when he could often have let them live, knowing their stricken aircraft were about to crash behind the German lines. His savagery became more pronounced after his head wound. His cruelty was inherent and was not owed to the war.

He was an overtly loving son and brother and warm in his friendships with his comrades.

He was the typical country gentleman of any Western European nationality, keen on field sports. Judging from his character, he would have taken ardently to game fishing and revelled in the fight to dominate marlin, tuna and shark. He would have enjoyed it honestly, not in the Hemingway fashion of bogus masculinity and an unquenchable urge to brag about it. Manfred thrived where, as Douglas Bader expressed it, 'the hot lead was flying', not hot air in the bar of the Paris Ritz.

He derived enormous pleasure from the decorations he received, but this was not purely from vanity, it was a professional soldier's way of assessing his own worth. He accumulated a staggering array of these. For centuries, Germany had been a geographical area, not a specific country. The name embraced numerous small states, each ruled by a king, prince or duke. In 1871 the German Empire was founded and the King of Prussia became emperor. The rulers of some of the states continued to bestow their own decorations. Manfred's Iron Cross, Second Class and First Class, and *Pour le Mérite*, known as the Knight's Cross of the Iron Cross, were followed by nine more German awards: the Order of the House of Hohenzollern, Order of the Royal House of Oldenberg, Saxony Military Order of St Henry, Griffon Cross, Hessen Order of Phillips, Saxe-Coburg-Gotha Duke Karl Edward Medal, Lippe Schaumberg Cross, Bremen Hanseatic Cross and Lübeck Hanseatic Cross. Austria-Hungary gave him the Order of the Holy Crown, Imperial Order of the Iron Crown and Military Service Cross. From Bulgaria he received the Order of Military Valour; and from Turkey, the Star of Gallipoli, Imtjaz Medal and Liakat Medal.

It gave him pride, pleasure and satisfaction to do useful work. Of his life as an observer: 'This was a wonderful time. Life in the air service was much like cavalry life. Every morning and afternoon I flew on reconnaissance, gathering valuable information'. A year later, 'I love my new occupation as a pilot. I do not think anything else can attract me in this war'.

Like Ball, he did not spare his parents' feelings. In October 1916: 'During the last four weeks we have lost five aeroplanes out of ten.'

On another occasion he informed his mother,

'One of my wings broke at a height of 300 metres and it was a miracle that I reached the ground without a mishap. On the same day my old squadron lost three aeroplanes. It is possible that they met with the same accidents.'

In other words, don't be surprised if it happens to me again.

His combat reports were not written in the third person, like Ball's. They were crisp and, in their own way, detached. One dated 20 December 1916 is typical.

'Time 1.45 p.m. over Noreuil. Vickers two-seater [it was not, it was an FE2b]. Engine, Beardmore No. 791. Occupants, pilot Lieutenant L.G. D'Arcy [who was, in fact, the observer; the pilot was Sub-Lieutenant R.C. Whiteside, RNAS, the aircraft belonged to 18 Squadron RFC], observer unknown, no identification disc. About 1.45 p.m. with four aircraft I attacked an enemy squadron at 3,000 metres over Noreuil. The English had not yet been attacked and were in loose formation. I therefore had a chance to attack the last machine. I was leading and no other German aeroplanes were to be seen. After the first attack the enemy's engine began to smoke and the observer had been wounded. The machine spiralled down in wide turns. I followed and fired at close range. It was later found that I had killed the pilots. The aeroplane crashed to the ground between Quéant and Lagnicourt.'

He had no taste for patrolling on his own; it was part of his nature and his training to be a leader of men and there was also the matter of prudence, of having at least one companion so that they could protect each other.

It has to be conceded that Boelcke was the originator of fighter tactics that remained valid until after the Korean War. Immelmann deserves credit for his collaboration in evolving the system of the basic pair, but it was Boelcke alone who developed this into the elemental finger-four and must also be acknowledged as the finest squadron leader of the war. In this, Manfred was equal to him in situational awareness and tactical sense, but whereas Boelcke often allowed the other pilots with him to shoot down the enemy while he guarded them, Manfred took first pick while the others protected him. Manfred has to be rated the outstanding fighter *Geschwader* leader because there were no German fighter Wings in Boelcke's time and nobody who succeeded Manfred during the seven months between his death and the end of the war showed any superiority. He has a further claim to this for laying down the first rules for flying in squadron, or greater, formations; but in Vs of five, not finger-fours.

He was also the best shot. As an inspiration to other fighter pilots he surpassed Boelcke because he was unique as the top scorer of the whole war. As for the value of his eighty victories, forty-eight were over two-seater reconnaissance or bomber types, for which he has often been derided because they were usually easier victims than single-seaters. In his favour is the fact that their reports, photographs, artillery direction or bombing were collectively more important to the conduct of the ground war than forty-eight fighters. With regard to the skill and courage involved in shooting them down, thirteen were various marks of BE2 and easy meat. The eight Sopwith Camel single-seat fighters he took out reflect much more credit on his competence and bravery.

In comparison with his contemporary British and French fighter pilots, he lacked the dash of Ball, McCudden and Collishaw and he was no better a leader than they or Mannock, nor did he have Mannock's sharp brain and planning ability. If Ball had not been so regardless of caution that he virtually threw his life away; if McCudden had not made an uncharacteristic fatal error of judgment; if Mannock and Collishaw had appeared on the scene earlier, any one of them could have exceeded Manfred's total. If Fonck had been a trifle less calculating, Nungesser and Navarre less wild, and Guynemer had not gone so long without a rest that he was not fit to fly

163

at all when he was killed, all of them were capable of surpassing him.

There is one highly significant difference between the conditions in which the fighter pilots of the two world wars fought. In the first war, the RFC was not issued with parachutes. The office-bound generals who took the decisions did not deign to explain their reason; which was the disgraceful one that pilots might be tempted to abandon a fight prematurely. The same ukase applied to the *Luftstreitkräfte* until it was rescinded in early 1918. Manfred was wearing one when he was killed. One wonders how many British or Germans who died would have survived to fight again if, knowing their aircraft must be lost, they were able to save themselves; and how many more victories they would have scored?

After the war the RAF continued flying wartime types, still without parachutes, for seven years. The United States Army Air Corps was already equipped with parachutes, so two RAF officers were sent to America to learn all about them. It was not until 1925 that the Parachute Test Unit was formed to visit squadrons and demonstrate how to bale out. The seats in wartime aeroplanes were modified, and those in new types suitably designed, to take a parachute pack. The first RAF pilot to use a 'chute was Pilot Officer Eric Pentland, when his Avro 504K got into an inverted spin in 1926. Many of the high-scoring pilots in World War 2 had to bale out; some, more than once; among them Lacey and Mölders.

The only loners in the second war were the night fighter and intruder crews. Day fighters flew at least in pairs, except in the early days over the desert, when there were so few of them. The only other singleton sorties were not for seeking combat but on weather reconnaissance, convoy patrol or high-altitude photographic reconnaissance.

In comparison with the greatest exponents of air fighting and of leading fighter formations in the Second World War or the Korean War, it has to be assumed that if Manfred had been of their generation, he would have been equally successful.

It is impossible to be didactic about who was 'the best' fighter pilot or fighter leader in either world war. All the most individually successful were on a par and so were the most prominent Wing leaders. Opportunity was the great decisive factor; if Johnnie Johnson and Pat Pattle had changed places, their scores would most likely have been reversed; if Douglas Bader and Sailor Malan had swopped places, their Wings would have been equally well led; the same can be said about any of the best German, French or British squadron or Wing leaders changing places. Obviously, some had keener eyesight than others or superior situational awareness, were better shots or better flyers, better at their relationships with the pilots whom they led. The only distinctions one can accede are to those who were the first to introduce some innovation. Even so, if Boelcke had not worked out that operating in pairs or fours was best, someone else would have – very likely, Manfred; if Mölders had not revived this system, another fighter leader would have; probably Malan, Tuck, Bader or Galland.

Manfred von Richthofen was just one of the most determined and dedicated of his generation, one of the most inspirational and gifted leaders and one of the very best shots. He is remembered because he was the top scorer in the Great War and for the gaudy livery that made first his squadron and then his whole Wing conspicuous.

Chapter 21

Manfred von Richthofen's Victories

Manfred's first two successes, against *l'Aviation Militaire*, were not credited to him because both aircraft went down behind the French lines and could not be confirmed by German witnesses. The confirmed victories and the fate of pilots and observers are given below.

No	Date	Aircraft	Fate of Pilot/crew
1	17. 9.16	FE2b	Both DOW
2	23. 9.16	Martinsyde G100	Pilot KIA
3	30. 9.16	FE2b	Both KIA
4	7.10.16	BE2	Pilot KIA
5	16.10.16	BE2	Pilot KIA
6	25.10.16	BE2	Pilot KIA
7	3.11.16	FE2b	Both KIA
8	9.11.16	BE2c	Pilot KIA no obs.
9	20.11.16	BE2c	Both POW
10	20.11.16	FE2b	Pilot POW obs. KIA
11	23.11.16	DH2	Pilot KIA
12	11.12.16	DH2	Pilot POW
13	20.12.16	DH2	Pilot KIA
14	20.12.16	FE2b	Both KIA
15	27.12.16	FE2b	Pilot WIA obs. POW
16	4. 1.17	Sopwith Pup	Pilot KIA
17	23. 1.17	RE8	Pilot KIA
18	24. 1.17	FE2b	Both WIA and POW

KIA=killed in action; DOW=died of wounds; WIA=wounded in action; POW=prisoner of war

19	1. 2.17	BE2d	Both DOW
20	14. 2.17	BE2d	Pilot WIA obs. KIA
21	14. 2.17	BE2c	Pilot WIA no obs.
22	4. 3.17	BE2d	Both WIA
23	4. 3.17	Sop. 1½-strutter	Both KIA
24	6. 3.17	BE2c	Both KIA
25	9. 3.17	DH2	Pilot KIA
26	11. 3.17	BE2d	Both KIA
27	17. 3.17	FE2b	Both KIA
28	17. 3.17	BE2c	Both KIA
29	21. 3.17	BE2f	Both KIA
30	24. 3.17	Spad 7	Pilot WIA and POW
31	25. 3.17	Nieuport 17	Pilot POW
32	2. 4.17	BE2d	Both KIA
33	2. 4.17	Sop. 1½-strutter	Pilot POW obs. KIA
34	3. 4.17	FE2d	Pilot WIA, POW obs. KIA
35	5. 4.17	FE2a	Both POW obs. WIA
36	5. 4.17	FE2a	Pilot WIA, POW obs. DOW
37	7. 4.17	Nieuport 17	Pilot KIA
38	8. 4.17	Sop. 1½-strutter	Pilot WIA, POW obs. KIA
39	8. 4.17	BE2c	Both WIA
40	11. 4.17	BE2c	Both WIA
41	13. 4.17	RE8	Both KIA
42	13. 4.17	FE2b	Both KIA
43	13. 4.17	FE2b	Both KIA
44	14. 4.17	Nieuport 17	Pilot POW
45	16. 4.17	BE2c	Pilot WIA obs. KIA
46	22. 4.17	FE2b	Pilot POW obs. KIA
47	23. 4.17	BE2f	Both KIA
48	28. 4.17	BE2c	Pilot DOW obs. WIA, POW
49	29. 4.17	Spad 7	Pilot KIA
50	29. 4.17	FE2b	Both KIA
51	29. 4.17	FE2c	Both KIA
52	29. 4.17	Sop. Triplane	Pilot KIA
53	18. 6.17	RE8	Both KIA
54	23. 6.17	Spad 7	Pilot Unknown
55	24. 6.17	DH4	Both KIA
56	25. 6.17	RE8	Both KIA
57	2. 7.17	RE8	Both KIA
58	16. 8.17	Nieuport 17	Pilot KIA
59	26. 8.17	Spad 7	Pilot KIA
60	1. 9.17	RE8	Pilot POW obs. KIA

61	3. 9.17	Sop. Pup	Pilot POW
62	23.11.17	DH5	Pilot WIA
63	30.11.17	SE5a	Pilot KIA
64	12. 3.18	FE2b	Both POW
65	13. 3.18	Sop. Camel	Pilot WIA, POW
66	18. 3.18	Sop. Camel	Pilot POW
67	24. 3.18	SE5a	Pilot KIA
68	25. 3.18	Sop. Camel	Pilot KIA
69	26. 3.18	Sop. Dolphin,	Sqdn & pilot unknown
70	26. 3.18	RE8	Both KIA
71	27. 3.18	Sop. Camel	Pilot WIA, POW
72	27. 3.18	FE2b	Sqdn & pilot unknown
73	27. 3.18	FE2b	Both KIA
74	28. 3.18	AWFK8	Both KIA
75	2. 4.18	RE8	Both KIA
76	6. 4.18	Sop. Camel	Pilot KIA
77	7. 4.18	SE5a	Sqdn & Pilot unknown
78	7. 4.18	Sop. Camel	Pilot POW
79	20. 4.18	Sop. Camel	Pilot KIA
80	20. 4.18	Sop. Camel	Pilot WIA, POW.

Summary

	Single-seater pilots	Two-seater pilots	Observers
KIA	19	25	32
DOW		3	3
WIA	7	9	6

In those two-seaters in which the pilot did not have a machine-gun, the observer, who did, was the obvious target.

Being a two-seater pilot was almost as lethal, because the aircraft were so much slower than the contemporary fighters.

BIBLIOGRAPHY

Arndt, A. Die Fliegertruppe im Weltkrieg, *Reichsarchiv*.

Bickers, R.L.T. *Ginger Lacey, Fighter Pilot*, Hale, 1962.
    The First Great Air War, Hodder & Stoughton, 1988.
    The *Battle of Britain*, Salamander, 1990.
    The Desert Air War, Leo Cooper, 1991.

Bishop, W. *Winged Warfare*, Bailey Brothers & Swinfen, 1925.

Bodenschatz, K. Jagd in Flanderns Himmel – Aus den sechzehen Kampfmonaten des Jagdgeschwaders Freiherr von Richthofen, Munich, 1935.

Boelcke, O. Hauptmann Boelcke Feldberichte, Gotha, 1916. *Centre Historique de l'Armée de l'Air*. La Grande Guerre dans le Ciel, and La Guerre Aérienne.

Clark, A. *Aces High*, Collins, 1973.

Collishaw, R. *Air Command*, Kimber, 1973.

Corsini, P. La Partecipazione degli Aviatori Inglesi alle Operazione sul Fronte Italiano durante la Grande Guerra e il Cinquantenario della Fondazione della RAF, *Ufficio Storico Aeronautico Militare*.

Mencarelli, I. (1) Francesco Barracca, (2) Fulco Ruffo de Calabria, *Ufficio Storico Aeronautica Militare*.

Crundall, E. *Fighter Pilot on the Western Front*, Kimber, 1975.

Degelow, C. *Germany's Last Knight of the Air*, Kimber, 1979.

Dudgeon, J. *Mick, the story of Major Edward Mannock VC DSO MC RFC RAF*, Hale, 1981.

Gibbons, F. *The Red Knight of Germany*, Cassell, 1932.

Gron, H. Die Organisation des Deutches Heeres im Weltkrieg. E.S. Mittler & Sohn, Berlin, 1923.

Hawker, T. *Hawker VC*, Mitre Press, 1965.

Immelmann, F. *Immelmann: The Eagle of Lille*, Hamilton, 1935.

Jackson, Robert, *Aces' Twilight*, Sphere, 1988.

Jones, H. *The War in the Air*, vols 2–5, HMSO & Hamish Hamilton, 1969.

Kilduff, P. *Richthofen, Beyond the Red Baron*, Arms & Armour, 1993.

Lee, A. *No Parachute*, Jarrold, 1967.

Liddell Hart, B. *History of the First World War*, Cassell, 1970.

Neumann, G. *German Air Force in the Great War*, Chivers, 1969.

Nowarra, H. *Von Richthofen and the Flying Circus. Luftfart Archiv*, Berlin.

Porret, D. Les 'As' Français de la Grande Guerre, Cedocar, 1983.

Porro, A. La Guerra nel Aria 1915–1918. *Edizione Mate*, Milan, 1965.

Raleigh, W. *The War in the Air*, vol 1. HMSO & Hamish Hamilton, 1969.

Richthofen, M. von, Der Rote Kampflieger, Berlin, 1917.

Richthofen, Kunigunde Freifrau von, Mein Kriegestagebuch, Berlin 1937.

Rickenbacker, E. *Fighting the Flying Circus*, Bailey Brothers & Swinfen, 1973.

'Theta', *War Flying*, John Murray, 1916.

Winter, D. *The First of the Few*, Allen Lane, 1982.

Woodhouse, J and Embleton, G. *The War in the Air 1914–1918*, Almark, 1934.

Zuerl, W. Pour le Mérite-Flieger, Curt Pechstein, Munich, 1938.

INDEX